Table of Contents

SOME ECLECTIC MATRIX THEORY

Kenneth S. Miller, Ph. D.
Riverside Research Institute
New York, New York 10036

Robert E. Krieger Publishing Company
Malabar, Florida
1987

Original Edition 1987

Printed and Published by
ROBERT E. KRIEGER PUBLISHING COMPANY, INC.
KRIEGER DRIVE
MALABAR FL 32950

Printed in the United States of America

Library of Congress Cataloging-in-Publication Data

Miller, Kenneth S.
Some eclectic matrix theory.

Bibliography: p.
1. Matrices. I. Title.
QA184.M55 1987 512.9′434 85–18454
ISBN 0–89874–895–X

10 9 8 7 6 5 4 3 2

Foreword

We live in a multidimensional world surrounded by an environment whose impact is described in terms of probability. Clearly, applied mathematics will be concerned with matrices and statistics and the theory of these topics is usually part of the general education of a user of mathematics. The logical development of these theories is "geometric" in the analysis sense of the word. One deals with linear spaces, vectors, transformations and eigenvectors, and the integral and tangential calculus on the number spaces of analysis. These are concepts developed in the long history of mathematics to hold our understanding.

But an applied mathematician has a commitment not only to understand a situation but to do something about it. In the present work, the geometric theory yields a set of formulas in matrix algebra which give the answer to specific problems, such as inverses. Clearly these formulas are of immediate practical value and their development requires both manipulative skill and extensions of the basic theory. These formulas also reflect the extensive experience of the author in utilizing matrix theory and statistics.

But the formulas are just one aspect of the metamorphism of mathematical theory into an effective tool for applications. In applied mathematics one seldom has the textbook association of problem and paradigm. One must deal with complicated situations, which, at best, are only approximately covered by past experience. What are needed are not only the various formulas but, rather, organized developments which represent the required guidance. The cogent matrix formulas indicate and outline these developments. Like stations on a subway map, they tell where to get on and where to get off but their association also indicates the branches of the system.

An example of this appears in the initial chapters of this book. If one is concerned with inverses, an opportunistic approach will seek out one eigenvalue. One needs both formulations of the desired situation and appropriate inverse formulas. Sections 1 to 4 provide these. There is an obvious sequence of formulas but, clearly, the full development has the real significance.

This book then represents the distilled essence of a considerable range of experience. There is an aphorism to the effect that those who do not study history are condemned to relive it. It is certainly more desirable to learn the available procedures than to try and construct them de novo.

At this moment there is the expectation that the new giant computers, which will soon be available to many users, will provide a significant increase in computing capability. Matrix concepts are, of course, the inspiration for the architecture of parallel computation and the broad range of experience in matrix theory and statistics will certainly constitute an important theoretical support for this development. It is true that whenever a new increase in computing capability appears, the opinion is expressed that one can utilize this power with simple primitive methods and ignore the previously developed skills and procedures. But one finds that this throws away a good deal of the new capacity and that computing power should be matched with sophisticated procedures.

Francis J. Murray

Preface

Many elementary properties of matrices that are useful in applied problems are not derived in a first course in matrix theory. For example, we mention the endless matrix identities [prove that $(\mathbf{F}^{-1}-\mathbf{G}^{-1})^{-1}=\mathbf{F}+\mathbf{F}(\mathbf{G}-\mathbf{F})^{-1}\mathbf{F}$], the problem of finding the inverse of a matrix when the matrix is of some special form [prove that $(\mathbf{I}+\mathbf{H})^{-1}=\mathbf{I}-(1+\operatorname{tr}\mathbf{H})^{-1}\mathbf{H}$ if $\mathbf{H}$ has rank one], the problem of "higher order" matrices (Kronecker products), the problem of differentiating a matrix with respect to a matrix—just to mention a few. The purpose of this monograph is to present a number of these results which the author has found useful in various applied problems that have come to his attention over the years. We hope that at least some of them also will prove of use to the reader. Most of the propositions require only the knowledge usually contained in a first course in linear algebra.

Before embarking on our eclectic trip, we recall a few useful and standard facts about matrices.

Let $\mathfrak{N}$ be the class of all $n \times n$ square matrices. If $\mathbf{A}, \mathbf{B} \in \mathfrak{N}$, then we shall say that $\mathbf{A}$ and $\mathbf{B}$ are *similar* (written $\mathbf{A} \sim \mathbf{B}$) if there exists a nonsingular matrix $\mathbf{P}$ such that

$$\mathbf{P}^{-1}\mathbf{A}\mathbf{P}=\mathbf{B} \quad . \tag{1}$$

Thus we may decompose $\mathfrak{N}$ into canonical sets (equivalence classes) such that if **A** and **B** are in a particular canonical set, $\mathbf{A} \sim \mathbf{B}$, and if **A** and **B** are not in the same canonical set, $\mathbf{A} \nsim \mathbf{B}$. In each canonical set $\mathcal{S}$ choose a representative element, say **J**. Then all matrices in $\mathcal{S}$ are similar to **J**. Generally it is convenient to let **J** be the Jordan normal form.

If **A** is symmetric, then its Jordan normal form **D** is diagonal, $\mathbf{D} = \|\lambda_i \delta_{ij}\|$ where the λ_i are the n (not necessarily distinct) eigenvalues of **A**. Thus there exists a nonsingular matrix **P** such that

$$\mathbf{P}^{-1}\mathbf{A}\mathbf{P} = \mathbf{D} \ . \tag{2}$$

In the symmetric case we may choose **P** to be *orthogonal* (that is, its transpose $\mathbf{P}'$ is equal to its inverse $\mathbf{P}^{-1}$) and (2) may be written as

$$\mathbf{P}'\mathbf{A}\mathbf{P} = \mathbf{D} \ . \tag{3}$$

Now suppose that **A** is positive definite. Then $\lambda_i > 0$ for all i, $1 \leqq i \leqq n$. Let $\mathbf{Q} = \|\lambda_i^{-\frac{1}{2}} \delta_{ij}\|$. Then from (3)

$$\mathbf{Q}'(\mathbf{P}'\mathbf{A}\mathbf{P})\mathbf{Q} = \mathbf{Q}'\mathbf{D}\mathbf{Q} = \mathbf{I} \tag{4}$$

where **I** is the identity matrix. If we let $\mathbf{R} = \mathbf{P}\mathbf{Q}$, then for **A** positive definite,

$$\mathbf{R}'\mathbf{A}\mathbf{R} = \mathbf{I} \ .$$

Or, letting $\mathbf{R}^{-1} = \mathbf{S}$, we see that any positive definite matrix **A** may be factored as

$$\mathbf{A} = \mathbf{S}'\mathbf{S} \ . \tag{5}$$

Naturally we hope that the reader has some acquaintance with the above concepts. Furthermore we shall assume familiarity with other fundamental matrix results including, but not limited to, such topics as the Hamilton-Cayley theorem and partitioned matrices. (For example, see [Birkhoff and MacLane], [Halmos], [Hamburger and Grimshaw], [Hohn].) The necessity to consider certain facets of some mathematical theories not directly related to linear algebra occasionally will arise. Appropriate appendices have been included to cover these situations.

A few remarks on notation are in order. We shall represent vectors and matrices by bold face symbols, as we have done above. Occasionally we shall use script letters to represent matrices. A prime on a matrix or vector always will denote the transpose. Unprimed vectors are always column vectors; thus primed vectors are always row vectors.

If in Section a we have occasion to refer to Theorem x of Section b (with a≠b), then we shall write Theorem b.x. Similar remarks apply to equations. Thus, for example, when in Section 13 we refer to equation (17) of Section 10, we shall call it equation (10.17).

K. S. M.

1. INTRODUCTION

The reader probably first was introduced to matrices via the problem of solving a system of linear algebraic equations such as

$$
\begin{aligned}
&a_{11}x_1+a_{12}x_2+\cdots+a_{1n}x_n=b_1 \\
&a_{21}x_1+a_{22}x_2+\cdots+a_{2n}x_n=b_2 \\
&\cdot \quad \cdot \quad \cdot \quad \cdot \quad \cdot \\
&a_{m1}x_1+a_{m2}x_2+\cdots+a_{mn}x_n=b_m \ .
\end{aligned}
\tag{1}
$$

The a_{ij}, $1 \leqq i \leqq m$, $1 \leqq j \leqq n$, and b_i, $1 \leqq i \leqq m$, were assumed to be known constants, and the problem was to find the indeterminates x_1, $x_2, \cdots, x_n$. If we let $\mathbf{x}'=\{x_1, \cdots ,x_n\}$ be an n-dimensional vector, $\mathbf{b}'=\{b_1, \cdots ,b_m\}$ an m-dimensional vector and

$$\mathbf{A}=\|a_{ij}\|=\begin{bmatrix} a_{11} & \cdots & a_{1n} \\ \cdot & \cdot & \cdot \\ a_{m1} & \cdots & a_{mn} \end{bmatrix}$$

an $m \times n$ dimensional matrix, then (1) may be written succinctly as

$$\mathbf{A}\mathbf{x}=\mathbf{b} \ . \tag{2}$$

Initially it was assumed that $m=n$ so that $\mathbf{A}$ was a square matrix. Then if $\mathbf{A}$ is nonsingular, the solution of (1) or (2) assumes the elegant form

$$\mathbf{x}=\mathbf{A}^{-1}\mathbf{b} \ . \tag{3}$$

This formula, however, tells us precious little about how to find $\mathbf{A}^{-1}$.

Now suppose $\mathbf{A}$ is $m \times n$ dimensional where m is not necessarily equal to n. Then we know that (2) has a solution only if the rank of $\mathbf{A}$ and the rank of the $m \times (n+1)$ dimensional augmented matrix $\mathbf{B}=[\mathbf{A}\ \mathbf{b}]$ are the same. If $\mathbf{A}$ and $\mathbf{B}$ have the same rank, say r, then we know that the values of certain $n-r$ of the x_i's may be assigned at pleasure, and the remaining x_i's will be determined uniquely.

Let us now give another—perhaps less familiar—illustration of how matrices are useful. What I have in mind is the problem of least squares or linear regression—a technique extensively used in statistical theory. Fortunately, the basic idea may be expounded without invoking any stochastic concepts.

We begin with a simple problem. Suppose at some value of time, say t_1, we measure a physical quantity, say z, and obtain the numerical value

z_1. For example, z could be the velocity of a projectile, the height of a growing plant, the temperature of a chemical reaction. If we repeat the experiment at n distinct values of time, say $t_1, \cdots, t_n$, we get n corresponding observations $z_1, \cdots, z_n$, and we may plot z as a function of t. This is the situation illustrated in Fig. 1 below (for $n=7$). Frequently the graph (such as Fig. 1) suggests that a straight line might be a good fit to the data over the time interval $[t_1, t_n]$. The problem we set ourselves is to find the best straight line.

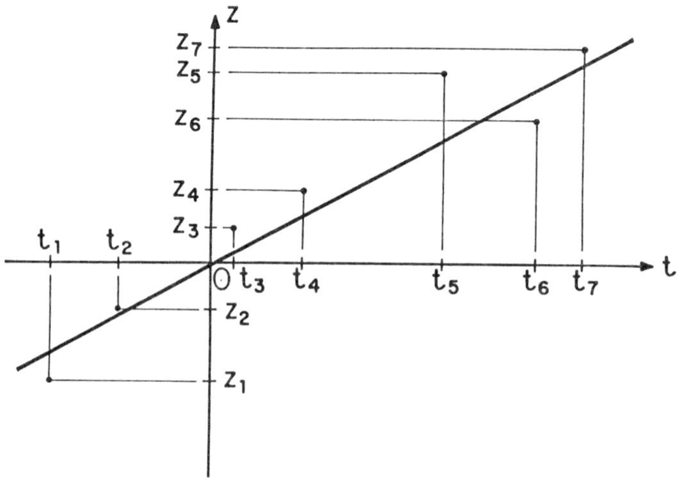

Figure 1

Now

$$z = x_1 + x_2 t \tag{4}$$

is the equation of a straight line. Thus all we have to do is to choose x_1 and x_2 so that (4) represents the best straight line approximation. But what do we mean by "best"? A commonly used criterion (which we too shall adopt) is the *least squares* fit. That is, we choose x_1 and x_2 so that the sum of the squares of the differences between the observations and the straight line is minimized. Analytically, this means that we must determine x_1 and x_2 so that

$$v = \sum_{i=1}^{n} [z_i - (x_1 + x_2 t_i)]^2 \tag{5}$$

is a minimum. If we set $\partial v/\partial x_1$ and $\partial v/\partial x_2$ equal to zero, we obtain the pair of linear simultaneous equations

$$\begin{aligned} nx_1 + (\Sigma\, t_i)x_2 &= \Sigma\, z_i \\ (\Sigma\, t_i)x_1 + (\Sigma\, t_i^2)x_2 &= \Sigma\, t_i z_i \end{aligned} \tag{6}$$

for determining x_1 and x_2. Of course we can solve these equations for x_1 and x_2 (provided $n \geqq 2$); but the result is far from elegant.

Let us now improve on the form of the solution by using matrix algebra. First we write (6) as

$$\begin{bmatrix} n & \Sigma\, t_i \\ \Sigma\, t_i & \Sigma\, t_i^2 \end{bmatrix} \begin{bmatrix} x_1 \\ x_2 \end{bmatrix} = \begin{bmatrix} \Sigma\, z_i \\ \Sigma\, t_i z_i \end{bmatrix} . \tag{7}$$

Now define the two-dimensional parameter vector $\mathbf{x}$ as

$$\mathbf{x}' = \{x_1, x_2\} ,$$

the n-dimensional observation vector $\mathbf{z}$ as

$$\mathbf{z}' = \{z_1, \cdots, z_n\} , \tag{8}$$

and the $n \times 2$ dimensional matrix of independent variables as

$$\mathbf{H} = \begin{bmatrix} 1 & t_1 \\ 1 & t_2 \\ \cdot & \cdot \\ 1 & t_n \end{bmatrix} . \tag{9}$$

Then we may write (6) or (7) compactly as

$$(\mathbf{H}'\mathbf{H})\mathbf{x} = \mathbf{H}'\mathbf{z} . \tag{10}$$

Since we have assumed the t_i's to be distinct (and tacitly assumed $n \geqq 2$), the matrix $\mathbf{H}'\mathbf{H}$ is nonsingular, and hence its inverse exists. If we call $\hat{\mathbf{x}}$ the solution of (10), then

$$\hat{\mathbf{x}} = (\mathbf{H}'\mathbf{H})^{-1}\mathbf{H}'\mathbf{z} . \tag{11}$$

Thus (11) represents the elegant matrix solution of our least squares problem.

How did we know how to choose $\mathbf{H}$ [see (9)]? To answer this question let us return to (5). We see that v is the square of the norm of a vector, say $\mathbf{v}' = \{v_1, \cdots, v_n\}$ whose ith component v_i is

$$v_i = z_i - (x_1 + x_2 t_i) , \quad 1 \leqq i \leqq n .$$

In vector form the above equations become

$$\mathbf{v}=\mathbf{z}-\mathbf{H}\mathbf{x} \tag{12}$$

where $\mathbf{H}$ is given by (9). Thus

$$v=|\mathbf{v}|^2=|\mathbf{z}-\mathbf{H}\mathbf{x}|^2 \ . \tag{13}$$

To be even more elegant, we shall see in Section 5 how to differentiate a scalar with respect to a vector. In this case we shall find that

$$\frac{\partial v}{\partial \mathbf{x}}=-2\mathbf{H}'\mathbf{z}+2\mathbf{H}'\mathbf{H}\mathbf{x} \ , \tag{14}$$

and setting $\partial v/\partial \mathbf{x}$ equal to the zero vector yields (10).

One might ask why we do not set $\mathbf{z}-\mathbf{H}\mathbf{x}$ equal to the zero vector. Then [see (13)] we would have $v=0$ and certainly v would be minimized. But as we have seen earlier, the equation $\mathbf{z}-\mathbf{H}\mathbf{x}=\mathbf{0}$ does not have a solution unless the ranks of $\mathbf{H}$ and $[\mathbf{H} \quad \mathbf{z}]$ are equal. If they *do* have the same rank, then (11) implies that $\hat{\mathbf{x}}=\mathbf{x}^*$ where $\mathbf{x}^*$ is the (unique) solution of $\mathbf{H}\mathbf{x}=\mathbf{z}$.

Note that the assumption that the independent variable t was "time" is unessential. In many applications t could have other interpretations. For example, if we were measuring the atmospheric pressure at different values of altitude, our independent variable would be spatial rather than temporal.

An immediate generalization of the above problem is to assume that the parameter vector $\mathbf{x}$ has p (instead of two) components, and that we wish to choose $x_1, \cdots, x_p$ such that

$$v=\sum_{i=1}^{n}[z_i-(h_{i1}x_1+h_{i2}x_2+\ \cdots\ +h_{ip}x_p)]^2 \tag{15}$$

is minimized. (The independent variables h_{ij} are assumed given—they play the role of 1 and t_i in the earlier problem. As before, $z_1, \cdots, z_n$ are the observations.) For the problem to be meaningful we must have at least as many observations as parameters, that is, $n \geqq p$. (Try to solve the simpler problem with $n=1$.)

If we let $\mathbf{x}$ be the p-dimensional parameter vector with components $x_1, \cdots, x_p$, and let

$$\mathbf{H}=\|h_{ij}\|=\begin{bmatrix} h_{11} & \cdots & h_{1p} \\ \cdot & \cdot & \cdot \\ h_{n1} & \cdots & h_{np} \end{bmatrix}$$

be the $n \times p$ dimensional matrix of independent variables, while $\mathbf{z}$ is still the n-dimensional vector of observations [see (8)], then (15) may be written as

$$v = |\mathbf{z} - \mathbf{H}\mathbf{x}|^2 \tag{16}$$

—which has the same form as (13). And if $\mathbf{H}'\mathbf{H}$ is nonsingular, the desired $\mathbf{x}$ is given by (11). (In all well posed practical problems $\mathbf{H}'\mathbf{H}$ is invariably nonsingular.)

In the two problems considered above, namely the solution of a system of simultaneous linear equations and the least squares fit, the answer involved the inversion of a matrix; see (3) and (11). The ubiquitous inverse appears in almost every nontrivial matrix problem. Hence it is of great importance to find techniques (other than the obvious brute force attacks) that allow us to find inverses. This recurring theme will permeate much of our work, (see Sections 4, 6, 11, 12, 14.)

We now are prepared to begin our eclectic trip. In Sections 2 and 3 we shall consider some useful matrix formulas. Section 4 will be our first attack on the inversion problem, and will include a very basic result based on the Sherman-Morrison formula. Next we shall introduce the concept of differentiating a scalar with respect to a vector, and in Section 6 we shall return to the matrix inversion theme. A few applications to positive definite matrices will appear in Section 7. Then we shall consider some neat results, with applications, involving matrices of rank one. Higher order matrices form the content of Section 9, followed by a treatment of elementary matrices useful in dealing with Kronecker products. We then have two more sections dealing with inverses of matrices. Next we extend the results of Section 5 to the problem of differentiating a matrix with respect to a matrix. In the penultimate section we consider some results associated with the inversion of matrices of special form; and in the final section we treat a potpourri of applications—mainly of the statistical variety.

2. ELEMENTARY SYMMETRIC FUNCTIONS

We wish to establish some preliminary results that will be useful later. A number of these theorems are most easily proved using the characteristic polynomial of a matrix. Thus it is fitting that we say a few words about certain elementary properties of polynomials.

The fundamental theorem of algebra states that any polynomial of the nth degree (with real or complex coefficients) has precisely n (real or complex) zeros. Thus if

$$P(x)=x^n-\sigma_1x^{n-1}+\cdots+(-1)^n\sigma_n \tag{1}$$

is a polynomial of the nth degree, and if $x_1,\cdots,x_n$ are the zeros of P, then we also may write $P(x)$ as

$$P(x)=\prod_{i=1}^{n}(x-x_i) \;.$$

On multiplying out the above expression and comparing the result with (1) we see that the σ_i may be expressed in terms of the x_j. Explicitly,

$$\sigma_0=1$$

$$\sigma_1=\sum_{i=1}^{n}x_i$$

$$\sigma_2=\frac{1}{2!}\sum_{i,j=1}^{n}{}'\,x_ix_j$$

$$\sigma_3=\frac{1}{3!}\sum_{i,j,k=1}^{n}{}'\,x_ix_jx_k$$

$$\cdot \qquad \cdot \qquad \cdot \qquad \cdot$$

$$\sigma_m=\frac{1}{m!}\sum_{i_1,i_2,\cdots,i_m=1}^{n}{}'\,x_{i_1}x_{i_2}\cdots x_{i_m}$$

$$\cdot \qquad \cdot \qquad \cdot \qquad \cdot$$

$$\sigma_n=x_1x_2\cdots x_n$$

$$\sigma_{n+r}=0 \;,\; r>0$$

where the prime on the summation sign means that we sum only over terms with distinct subscripts. We recall that $\sigma_1,\cdots,\sigma_n$ are called the elementary symmetric functions of $x_1, \cdots, x_n$, [Weisner, p. 105], [van der Waerden, p. 78].

Now every symmetric function may be expressed in terms of elementary symmetric functions. In particular we wish to consider sums of powers. Let

$$s_m = \sum_{i=1}^{n} x_i^m \ , \ m = 1,2, \cdots \ . \tag{2}$$

The recursive formulas

$$s_m = \sum_{k=1}^{m-1} (-1)^{k+1} \sigma_k s_{m-k} + (-1)^{m+1} m \sigma_m \ , \ m = 1,2, \cdots \tag{3}$$

and

$$\sigma_m = \frac{1}{m} \sum_{k=1}^{m} (-1)^{k+1} s_k \sigma_{m-k} \ , \ m = 1,2, \cdots \tag{4}$$

are known as *Newton's identities* [Weisner, p. 115]. They yield an algorithm for explicitly expressing the s_m's as a polynomial in the σ_k's, and conversely. Let us write

$$\sigma_m = \phi_m(s_1, \cdots, s_m) \ , \ m = 1,2, \cdots \ . \tag{5}$$

Then ϕ_m is a polynomial. For example,

$$\begin{aligned}
\phi_1(s_1) &= s_1 \\
\phi_2(s_1,s_2) &= \frac{1}{2!}(s_1^2 - s_2) \\
\phi_3(s_1,s_2,s_3) &= \frac{1}{3!}(s_1^3 - 3s_1s_2 + 2s_3) \\
\phi_4(s_1,s_2,s_3,s_4) &= \frac{1}{4!}(s_1^4 - 6s_1^2s_2 + 8s_1s_3 + 3s_2^2 - 6s_4) \\
\phi_5(s_1,s_2,s_3,s_4,s_5) &= \frac{1}{5!}(s_1^5 - 10s_1^3s_2 + 15s_1s_2^2 + 20s_1^2s_3 \\
&\quad - 20s_2s_3 - 30s_1s_4 + 24s_5) \\
\phi_6(s_1,s_2,s_3,s_4,s_5,s_6) &= \frac{1}{6!}(s_1^6 + 45s_1^2s_2^2 - 15s_1^4s_2 - 15s_2^3 + 40s_1^3s_3 \\
&\quad - 120s_1s_2s_3 + 40s_3^2 - 90s_1^2s_4 + 90s_2s_4 \\
&\quad + 144s_1s_5 - 120s_6) \ .
\end{aligned} \tag{6}$$

From the above arguments we readily deduce:

Theorem 1. Let p be a polynomial of degree n with zeros $x_1, \cdots ,x_n$. Let q be a polynomial of degree r with zeros $\xi_1, \cdots ,\xi_r$. If $s_m=t_m$, $m=1,2, \cdots$ where s_m is the sum of the mth powers of the x_i and t_m is the sum of the mth powers of the ξ_i; then the nonzero zeros of p and q are identical.

Proof. If we write

$$p(x)=x^n-\sigma_1x^{n-1}+ \cdots + (-1)^n\sigma_n$$

$$q(x)=x^r-\tau_1x^{r-1}+ \cdots +(-1)^r\tau_r$$

and

$$s_m=\sum_{i=1}^{m} x_i^m \ , \ m=1,2, \cdots$$

$$t_m=\sum_{i=1}^{m} \xi_i^m \ , \ m=1,2, \cdots$$

then [see (5)]

$$\sigma_m=\phi_m(s_1, \cdots ,s_m)$$

$$\tau_m=\phi_m(t_1, \cdots ,t_m) \ .$$

But by hypothesis $s_m=t_m$, $m=1,2, \cdots$. Thus $\sigma_m=\tau_m$ for all positive m. ■

Now let **H** be an $n\times n$ matrix. If **J** is the Jordan normal form of **H**, and **I** the identity matrix, then

$$\det (\mathbf{H}^m-\lambda\mathbf{I})=\det (\mathbf{J}^m-\lambda\mathbf{I})$$

for any positive integer m. Thus if $\lambda_1, \cdots ,\lambda_n$ are the eigenvalues of **H**, we see that $\lambda_1^m, \cdots , \lambda_n^m$ are the eigenvalues of $\mathbf{H}^m$. In particular,

$$\text{tr } \mathbf{H}^m=\sum_{i=1}^{n} \lambda_i^m \ , \ m=1,2, \cdots \ . \tag{7}$$

Using this result we may prove:

Theorem 2. Let **H** be a square matrix and p its characteristic polynomial. Then the coefficients of λ in $p(\lambda)$ are polynomials in traces of powers of **H**.

Proof. Let **H** be an $n\times n$ matrix and let

$$p(\lambda)=\lambda^n-\sigma_1\lambda^{n-1}+ \cdots +(-1)^n\sigma_n$$

be its characteristic polynomial. From (5),

$$\sigma_m = \phi_m(s_1, \cdots, s_m)$$

where ϕ_m is a polynomial,

$$s_k = \sum_{i=1}^{k} \lambda_i^k \ , \ k = 1,2, \cdots$$

and the λ_i are the eigenvalues of **H**. By (7), $s_k = \text{tr } \mathbf{H}^k$. ■

From Theorem 1 and the relationship expressed by (7) we have the immediate corollaries:

Corollary 1. Let **H** and **K** be square matrices with the property that tr $\mathbf{H}^m = \text{tr } \mathbf{K}^m$ for $m = 1,2, \cdots$. Then the nonzero eigenvalues of **H** and **K** are identical.

Observe the **H** and **K** need not have the same dimension.

Corollary 2. Let **A** be an $n \times r$ matrix and **B** an $r \times n$ matrix. Then the nonzero eigenvalues of **AB** and **BA** are identical.

Proof. It suffices to note that tr $(\mathbf{AB})^m = \text{tr } (\mathbf{BA})^m$ for all positive integers m and then invoke Corollary 1. ■

Corollary 3. Let **A** and **B** be as in Corollary 2. Then $\mathbf{I} + \mathbf{AB}$ is nonsingular if and only if $\mathbf{I} + \mathbf{BA}$ is nonsingular.

Proof. If $\mathbf{I} + \mathbf{AB}$ is nonsingular, -1 cannot be an eigenvalue of **AB**; and since by Corollary 2, **AB** and **BA** have the same nonzero eigenvalues, -1 cannot be an eigenvalue of **BA**. Thus det $(\mathbf{I} + \mathbf{BA}) \neq 0$. ■

If **H** is a square matrix with eigenvalues λ_i, then

$$\det \mathbf{H} = \prod_i \lambda_i \ .$$

This immediately leads to:

Theorem 3. Let **H** be an $n \times n$ square matrix. Let $\sigma_1, \cdots, \sigma_n$ be the elementary symmetric functions of the eigenvalues of **H**. Then

$$\det \mathbf{H} = \sigma_n \ .$$

3. SOME MATRIX IDENTITIES

We now shall prove a sequence of matrix identities leading up to the Sherman-Morrison formula. In the next section this important result will be our starting point in proving some very basic theorems in the theory of matrix inversion.

Lemma 1. Let $\mathbf{A}$ be an $n \times r$ matrix and $\mathbf{B}$ an $r \times n$ matrix. Let $\mathbf{I}+\mathbf{BA}$ be nonsingular. Then

$$(\mathbf{I}+\mathbf{AB})^{-1}\mathbf{A}=\mathbf{A}(\mathbf{I}+\mathbf{BA})^{-1} \tag{1}$$

and

$$(\mathbf{I}+\mathbf{AB})^{-1}=\mathbf{I}-(\mathbf{I}+\mathbf{AB})^{-1}\mathbf{AB} \ . \tag{2}$$

Proof. Consider the identity

$$\mathbf{A}(\mathbf{I}+\mathbf{BA})=(\mathbf{I}+\mathbf{AB})\mathbf{A} \ . \tag{3}$$

By hypothesis, $\mathbf{I}+\mathbf{BA}$ is nonsingular, and by Corollary 2.3, so is $\mathbf{I}+\mathbf{AB}$. Multiplying (3) on the right and left by the inverses of these matrices yields (1).

To prove (2), we start with the identity

$$\begin{aligned}\mathbf{I}&=(\mathbf{I}+\mathbf{AB})^{-1}(\mathbf{I}+\mathbf{AB})\\&=(\mathbf{I}+\mathbf{AB})^{-1}+(\mathbf{I}+\mathbf{AB})^{-1}\mathbf{AB}\end{aligned}$$

and rearrange terms. ■

Lemma 2. If $\mathbf{F}$ and $\mathbf{G}$ are nonsingular square matrices and $\mathbf{G}-\mathbf{F}$ also is nonsingular, then $\mathbf{F}^{-1}-\mathbf{G}^{-1}$ is nonsingular and

$$(\mathbf{F}^{-1}-\mathbf{G}^{-1})^{-1}=\mathbf{F}+\mathbf{F}(\mathbf{G}-\mathbf{F})^{-1}\mathbf{F} \ . \tag{4}$$

Furthermore, if $\mathbf{F}$ and $\mathbf{G}-\mathbf{F}$ are positive definite, then

$$\mathbf{F}^{-1}-\mathbf{G}^{-1} \tag{5}$$

also is positive definite.

Proof. In Lemma 1 let $\mathbf{A}=\mathbf{G}^{-1}$ and $\mathbf{B}=-\mathbf{F}$. Then $\mathbf{I}+\mathbf{BA}=\mathbf{I}-\mathbf{FG}^{-1}=(\mathbf{G}-\mathbf{F})\mathbf{G}^{-1}$. Since $\mathbf{G}-\mathbf{F}$ is nonsingular by hypothesis, we infer that $\mathbf{I}+\mathbf{BA}$ is nonsingular. Thus (2) implies (4).

If $\mathbf{F}$ and $\mathbf{G}-\mathbf{F}$ are positive definite, then the right-hand side of (4) also is positive definite. Hence (5) is positive definite. ■

A consequence of these identities is *Woodbury's formula* [Woodbury].

Woodbury's Formula. Under the hypotheses of Lemma 1,

$$(\mathbf{I}+\mathbf{A}\mathbf{B})^{-1}=\mathbf{I}-\mathbf{A}(\mathbf{I}+\mathbf{B}\mathbf{A})^{-1}\mathbf{B} \ . \tag{6}$$

Proof. Substitute (1) on the right-hand side of (2). ■

Let $\mathbf{C}$ be an $n\times n$ nonsingular matrix. Then by the transformation

$$\mathbf{U}=\mathbf{C}^{-1}\mathbf{A}$$

$$\mathbf{V}=\mathbf{B}\mathbf{C}^{2}$$

we obtain the following generalization of (6),

$$(\mathbf{C}+\mathbf{U}\mathbf{V})^{-1}=\mathbf{C}^{-1}-\mathbf{C}^{-1}\mathbf{U}(\mathbf{I}+\mathbf{V}\mathbf{C}^{-1}\mathbf{U})^{-1}\mathbf{V}\mathbf{C}^{-1} \ . \tag{7}$$

If we let $r=1$, then $\mathbf{A}$ becomes a column vector, say $\mathbf{u}$, and $\mathbf{B}$ becomes a row vector, say $\mathbf{v}'$. In this case, Woodbury's formula reduces to the *Sherman-Morrison formula* [Sherman and Morrison]

$$(\mathbf{I}+\mathbf{u}\mathbf{v}')^{-1}=\mathbf{I}-\frac{1}{1+\mathbf{v}'\mathbf{u}}\,\mathbf{u}\mathbf{v}' \tag{8}$$

(since $\mathbf{v}'\mathbf{u}$ is a scalar).

4. MATRIX INVERSION I—Inverse of the sum of matrices

The Sherman-Morrison formula is fundamental in our attack on the matrix inversion problem. Let $\mathbf{H}$ be a square matrix of rank one. Then there exist nonzero vectors $\mathbf{u}$ and $\mathbf{v}$ such that $\mathbf{H}=\mathbf{u}\mathbf{v}'$ and conversely (see [Halmos]). Hence we may write equation (8) of Section 3 as

$$(\mathbf{I}+\mathbf{H})^{-1}=\mathbf{I}-\frac{1}{1+\operatorname{tr}\mathbf{H}}\mathbf{H} \qquad (1)$$

since $\mathbf{v}'\mathbf{u}=\operatorname{tr}\mathbf{v}'\mathbf{u}=\operatorname{tr}\mathbf{u}\mathbf{v}'=\operatorname{tr}\mathbf{H}$.

There is a fundamental distinction between (1) and (3.8), namely, the vectors $\mathbf{u}$ and $\mathbf{v}$ do not explicitly appear in (1). While it is true that any square matrix $\mathbf{H}$ of rank one may be written as the outer product of two vectors, and while it is true that the proof of (1) uses these vectors, the vectors themselves do not appear in the final formula.

Because of the importance of (1) we state it as a theorem and give a direct proof [Miller, 1981].

Theorem 1. Let $\mathbf{H}$ be an $n\times n$ matrix of rank one. Let $\mathbf{I}+\mathbf{H}$ be nonsingular. Then $1+\operatorname{tr}\mathbf{H}\neq 0$ and (1) holds.

Proof. Since $\mathbf{H}$ has rank one, at least $n-1$ eigenvalues of $\mathbf{H}$ must be zero. Since the sum of all the eigenvalues of $\mathbf{H}$ is equal to the trace of $\mathbf{H}$, the remaining eigenvalue must be tr $\mathbf{H}$. But

$$\det(\mathbf{I}+\mathbf{H})=1+\operatorname{tr}\mathbf{H} \ ,$$

and since $\mathbf{I}+\mathbf{H}$ is nonsingular by hypothesis, $1+\operatorname{tr}\mathbf{H}\neq 0$.

Now let $\mathbf{u}$ and $\mathbf{v}$ be nonzero vectors such that $\mathbf{H}=\mathbf{u}\mathbf{v}'$. Then

$$\mathbf{H}^2=\mathbf{u}\mathbf{v}'\mathbf{u}\mathbf{v}'=\mathbf{u}(\mathbf{v}'\mathbf{u})\mathbf{v}'=(\mathbf{v}'\mathbf{u})(\mathbf{u}\mathbf{v}')=(\operatorname{tr}\mathbf{H})\mathbf{H} \ ,$$

and multiplying (1) by $\mathbf{I}+\mathbf{H}$ we verify that the right-hand side of (1) is indeed the inverse of $\mathbf{I}+\mathbf{H}$. ■

A thinly disguised generalization of the above theorem is given in the corollary below.

Corollary 1. Let $\mathbf{G}$ and $\mathbf{G}+\mathbf{H}$ be nonsingular matrices and let $\mathbf{H}$ have rank one. Let $g = \operatorname{tr} \mathbf{H}\mathbf{G}^{-1}$. Then $g \neq -1$ and

$$(\mathbf{G}+\mathbf{H})^{-1} = \mathbf{G}^{-1} - \frac{1}{1+g}\mathbf{G}^{-1}\mathbf{H}\mathbf{G}^{-1} .$$

If $\mathbf{H}$ has rank zero, which implies that $\mathbf{H}=\mathbf{0}$, then $g=0$ and Corollary 1 is trivially true (as, of course, is Theorem 1).

We give an application of the above results. Suppose

$$\mathbf{\Lambda} = \begin{bmatrix} \lambda & 1 & \cdots & 1 \\ 1 & \lambda & \cdots & 1 \\ \cdot & \cdot & \cdot & \cdot \\ 1 & 1 & \cdots & \lambda \end{bmatrix}$$

is an $n \times n$ matrix with λ on its main diagonal and ones everywhere else. For values of λ for which $\mathbf{\Lambda}$ is nonsingular we wish to find the inverse of $\mathbf{\Lambda}$.

To apply Theorem 1 we write $\mathbf{\Lambda}$ in the form

$$\mathbf{\Lambda} = (\lambda-1)\mathbf{I} + \mathbf{\Delta}$$

where $\mathbf{I}$ is the $n \times n$ identity matrix and $\mathbf{\Delta}$ is an $n \times n$ matrix with ones everywhere. We see that $\mathbf{\Delta}$ has rank one. Thus if we let

$$\mathbf{\Lambda} = \mathbf{G} + \mathbf{H} \tag{2}$$

where $\mathbf{G} = (\lambda-1)\mathbf{I}$ and $\mathbf{H} = \mathbf{\Delta}$, we may apply Corollary 1 to obtain the inverse of $\mathbf{\Lambda}$, viz.:

$$\mathbf{\Lambda}^{-1} = \frac{1}{\lambda-1}\mathbf{I} - \frac{1}{(\lambda-1)(\lambda+n-1)}\mathbf{\Delta} .$$

In expanded form

$$\mathbf{\Lambda}^{-1} = \frac{1}{(\lambda-1)(\lambda+n-1)} \begin{bmatrix} \lambda+n-2 & -1 & \cdots & -1 \\ -1 & \lambda+n-2 & \cdots & -1 \\ \cdot & \cdot & \cdot & \cdot \\ -1 & -1 & \cdots & \lambda+n-2 \end{bmatrix} .$$

That is, $\mathbf{\Lambda}^{-1}$ is the scalar $(\lambda-1)^{-1}(\lambda+n-1)^{-1}$ times a matrix with $\lambda+n-2$ on its main diagonal and -1 everywhere else. (Observe that $\mathbf{\Lambda}$ will be nonsingular provided $\lambda \neq 1$ or $1-n$.)

In terms of the elementary symmetric functions σ_k of the eigenvalues of **H** we may write (1) in the form

$$(\mathbf{I}+\mathbf{H})^{-1}=\mathbf{I}-\frac{1}{1+\sigma_1}\mathbf{H} \tag{3}$$

where

$$\sigma_1=s_1=\operatorname{tr}\mathbf{H}$$

[see (2.5), (2.6), (2.7)]. Somewhat similar arguments lead to the formula

$$(\mathbf{I}+\mathbf{H})^{-1}=\mathbf{I}-\frac{1}{1+\sigma_1+\sigma_2}[(1+\sigma_1)\mathbf{H}-\mathbf{H}^2] \tag{4}$$

where **H** has rank two and $\mathbf{I}+\mathbf{H}$ is assumed to be nonsingular. Again from (2.5), (2.6) and (2.7) we see that

$$\sigma_1=s_1=\operatorname{tr}\mathbf{H}$$

(as above) and

$$\sigma_2=\tfrac{1}{2}(s_1^2-s_2)=\tfrac{1}{2}[(\operatorname{tr}\mathbf{H})^2-\operatorname{tr}\mathbf{H}^2]\ .$$

[Equation (4) also is valid if **H** has rank one or zero.] The above results suggest the following generalization which expresses the inverse of the sum of matrices as a finite sum of matrices.

Theorem 2. Let **H** be an $n\times n$ matrix of rank r. Let $\sigma_1, \cdots, \sigma_n$ be the elementary symmetric functions of the eigenvalues of **H**. If $\mathbf{I}+\mathbf{H}$ is nonsingular, then $A_r\neq 0$ and

$$(\mathbf{I}+\mathbf{H})^{-1}=\mathbf{I}-\frac{1}{A_r}[A_{r-1}\mathbf{H}-A_{r-2}\mathbf{H}^2+\cdots+(-1)^{r-1}A_0\mathbf{H}^r] \tag{5}$$

where $A_0=1$ and

$$A_m=1+\sum_{k=1}^{m}\sigma_k\ ,\quad m=1,2,\cdots,r\ . \tag{6}$$

Proof. Since **H** has rank r, at least $n-r$ of the eigenvalues of **H** are zero. Let $\lambda_1, \cdots, \lambda_r$ be the remaining eigenvalues. Then the minimal polynomial of **H** divides $\lambda p(\lambda)$ where

$$p(\lambda)=\lambda^r-\sigma_1\lambda^{r-1}+\sigma_2\lambda^{r-2}+\cdots+(-1)^r\sigma_r\ .$$

Since $\mathbf{I}+\mathbf{H}$ is nonsingular, -1 cannot be an eigenvalue of **H**. Therefore

$p(-1) \neq 0$. But

$$(-1)^r p(-1) = 1 + \sum_{k=1}^{r} \sigma_k = A_r \ .$$

Thus $A_r \neq 0$. In fact we see that $A_r = \det(\mathbf{I}+\mathbf{H})$ since $\det(\mathbf{H}-\lambda\mathbf{I}) = (-1)^n \lambda^{n-r} p(\lambda)$ implies

$$\det(\mathbf{I}+\mathbf{H}) = (-1)^n (-1)^{n-r} p(-1) = A_r \ .$$

Now if we multiply (5) by $\mathbf{I}+\mathbf{H}$ and use the fact that $\mathbf{H}p(\mathbf{H}) = \mathbf{0}$, we verify that the right-hand side of (5) is indeed the inverse of $\mathbf{I}+\mathbf{H}$. ∎

Since

$$\sigma_m = \phi_m(s_1, \cdots, s_m) \ , \ m = 1, 2, \cdots$$

where ϕ_m is a polynomial and $s_k = \operatorname{tr} \mathbf{H}^k$ [see (2.5) and (2.7)] we see that $(\mathbf{I}+\mathbf{H})^{-1}$ is a polynomial in $\mathbf{H}$ of degree r whose coefficients are rational functions in traces of powers of $\mathbf{H}$.

We sometimes shall write $\partial\mathbf{G} = n$ to express the fact that $\mathbf{G}$ is an $n \times n$ dimensional square matrix. We also shall use the notation $\rho\mathbf{A}$ to indicate the rank of a (not necessarily square) matrix $\mathbf{A}$.

5. THE DERIVATIVE OF A SCALAR

Let $\mathbf{F} = \|f_{ij}\|$ be a matrix. If for every i and j, f_{ij} is a differentiable function of x for x in some interval I, then we define the derivative of $\mathbf{F}$ as the matrix

$$\frac{d}{dx}\mathbf{F} = \left\| \frac{d}{dx} f_{ij} \right\|$$

for all $x \in I$. In other words, we have differentiated a matrix with respect to a scalar. This, of course, is nothing new. Now in Section 1 we used the concept of the derivative of a scalar with respect to a vector to obtain the least squares fit to a set of observations [see (1.14)]. Let us define this operation.

Suppose $\mathbf{x}' = \{x_1, \cdots, x_n\}$ is a vector and $s(\mathbf{x}) = s(x_1, \cdots, x_n)$ is a scalar function of $\mathbf{x}$. For example, the norm of $\mathbf{x}$ is a scalar function of $\mathbf{x}$. If $s(x_1, \cdots, x_n)$ is a differentiable function of its arguments for $\mathbf{x}$ in some region $\mathcal{D}$ of n-dimensional Euclidean space, then we define the derivative $ds/d\mathbf{x}$ for all $\mathbf{x} \in \mathcal{D}$ as

$$\frac{d}{d\mathbf{x}} s(\mathbf{x}) = \begin{bmatrix} \dfrac{\partial s}{\partial x_1} \\ \dfrac{\partial s}{\partial x_2} \\ \cdot \\ \cdot \\ \cdot \\ \dfrac{\partial s}{\partial x_n} \end{bmatrix}.$$

Thus the derivative of a scalar with respect to a (column) vector is defined as a (column) vector of the partial derivatives of s. The derivative of a scalar with respect to a row vector is

$$\frac{d}{d\mathbf{x}'} s(\mathbf{x}) = \left\{ \frac{\partial s}{\partial x_1}, \cdots, \frac{\partial s}{\partial x_n} \right\}.$$

Analogously we also could define the derivative of a scalar with respect to a matrix. However, for the present, we need only the above definitions. In Section 13 we shall consider the problem of differentiating a matrix with respect to a matrix. Then, of course, all the above definitions will be special cases.

Let us give a few simple examples. Suppose $\mathbf{a}' = \{a_1, \cdots, a_n\}$ is a vector independent of $\mathbf{x}$. We shall find the derivative of the inner product $\mathbf{a}'\mathbf{x}$ of $\mathbf{a}$ and $\mathbf{x}$ with respect to $\mathbf{x}$. In expanded form

$$\mathbf{a}'\mathbf{x} = a_1 x_1 + \cdots + a_n x_n$$

and

$$\frac{\partial}{\partial x_i} \mathbf{a}'\mathbf{x} = a_i \ , \quad 1 \leqq i \leqq n \ .$$

Thus by the definition of the derivative of a scalar with respect to a vector,

$$\frac{d}{d\mathbf{x}} \mathbf{a}'\mathbf{x} = \mathbf{a} \ . \tag{1}$$

The Euclidean norm $|\mathbf{x}|$ of $\mathbf{x}$ is

$$|\mathbf{x}| = (x_1^2 + \cdots + x_n^2)^{\frac{1}{2}}$$

and

$$\frac{\partial}{\partial x_i} |\mathbf{x}| = \frac{1}{|\mathbf{x}|} x_i \ , \quad 1 \leqq i \leqq n \ . \tag{2}$$

Thus

$$\frac{d}{d\mathbf{x}} |\mathbf{x}| = \frac{1}{|\mathbf{x}|} \mathbf{x} \ .$$

We leave it to the reader to show that

$$\frac{d}{d\mathbf{x}} \mathbf{x}'\mathbf{Q}\mathbf{x} = (\mathbf{Q} + \mathbf{Q}')\mathbf{x} \tag{3}$$

where $\mathbf{Q}$ is an $n \times n$ dimensional matrix independent of $\mathbf{x}$. In particular, if $\mathbf{Q}$ is symmetric, we have $\mathbf{Q} = \mathbf{Q}'$ and the above formula reduces to

$$\frac{d}{d\mathbf{x}} \mathbf{x}'\mathbf{Q}\mathbf{x} = 2\mathbf{Q}\mathbf{x} \ . \tag{3a}$$

(See [Miller, 1980]). Applying the above formulas to equation (13) of Section 1 leads to (1.14).

We turn now to a more difficult problem. Suppose we have an n-dimensional vector of observations $\mathbf{z}$, an $n \times p$ dimensional matrix $\mathbf{H}$ of independent variables with rank p, and a p-dimensional vector parameter $\mathbf{x}$. Now, as in Section 1, the "best" $\mathbf{x}$ is obtained by minimizing

$$v = |\mathbf{z} - \mathbf{H}\mathbf{x}|^2$$

and leads to the formula

$$\hat{\mathbf{x}} = (\mathbf{H}'\mathbf{H})^{-1}\mathbf{H}'\mathbf{z} \ .$$

So far, this is just the problem we solved in Section 1. Now suppose we impose the condition that the "best" $\mathbf{x}$ must be chosen from a set of vectors $\mathbf{x}$ which satisfy the condition

$$\mathbf{B}\mathbf{x} = \mathbf{0} \tag{4}$$

where $\mathbf{B}$ is an $r \times p$ matrix of rank r. That is, we must minimize $|\mathbf{z} - \mathbf{H}\mathbf{x}|^2$ subject to the constraint $\mathbf{B}\mathbf{x} = \mathbf{0}$. Problems of this nature arise, for example, in the theory of linear hypotheses (see [Lehmann]).

To solve this problem we introduce the vector Lagrange multiplier $\boldsymbol{\lambda}$ (see Appendix B). Then the problem becomes one of choosing $\mathbf{x}$ such that

$$w = |\mathbf{z} - \mathbf{H}\mathbf{x}|^2 - 2(\mathbf{B}\mathbf{x})'\boldsymbol{\lambda} \tag{5}$$

is minimum. (The "2" factor in the second term on the right-hand side of (5) is merely for convenience.) Applying the earlier differentiation formulas to (5) leads to

$$\frac{\partial w}{\partial \mathbf{x}} = -2\mathbf{H}'\mathbf{z} + 2\mathbf{H}'\mathbf{H}\mathbf{x} - 2\mathbf{B}'\boldsymbol{\lambda}$$

and setting this derivative equal to the zero vector yields

$$\mathbf{x} = (\mathbf{H}'\mathbf{H})^{-1}(\mathbf{B}'\boldsymbol{\lambda} + \mathbf{H}'\mathbf{z}) \ . \tag{6}$$

But $\mathbf{B}\mathbf{x} = \mathbf{0}$ [see (4)]. Thus we have, on multiplying (6) on the left by $\mathbf{B}$,

$$\mathbf{0} = \mathbf{B}(\mathbf{H}'\mathbf{H})^{-1}(\mathbf{B}'\boldsymbol{\lambda} + \mathbf{H}'\mathbf{z}) \ .$$

Now solve for $\boldsymbol{\lambda}$:

$$\boldsymbol{\lambda} = -[\mathbf{B}(\mathbf{H}'\mathbf{H})^{-1}\mathbf{B}']^{-1}\mathbf{B}(\mathbf{H}'\mathbf{H})^{-1}\mathbf{H}'\mathbf{z} \ .$$

Substituting in (6) then yields our constrained least squares fit

$$\hat{\hat{\mathbf{x}}} = (\mathbf{H}'\mathbf{H})^{-1}\mathbf{H}'\mathbf{z} - (\mathbf{H}'\mathbf{H})^{-1}\mathbf{B}'[\mathbf{B}(\mathbf{H}'\mathbf{H})^{-1}\mathbf{B}']^{-1}\mathbf{B}(\mathbf{H}'\mathbf{H})^{-1}\mathbf{H}'\mathbf{z} \ . \tag{7}$$

If the reader now has any doubts as to the efficacy of matrix notation, try to solve the above problem *not* using matrices!

Let us now apply the concept of differentiation with respect to a vector to an entirely different type of problem: namely, to the problem of finding the norm of a matrix.

The (Hilbert) norm $|\mathbf{G}|$ of a square matrix $\mathbf{G}$ may be defined as

$$|\mathbf{G}| = \max_{|\mathbf{x}|=1} |\mathbf{G}\mathbf{x}| \ . \tag{8}$$

We recall a few elementary properties of the norm. If a is an arbitrary scalar,

$$|a\mathbf{G}| = \max_{|\mathbf{x}|=1} |a\mathbf{G}\mathbf{x}| = |a| \max_{|\mathbf{x}|=1} |\mathbf{G}\mathbf{x}| \ .$$

Thus

$$|a\mathbf{G}| = |a| \ |\mathbf{G}| \ .$$

And for any vector $\mathbf{y}$,

$$|\mathbf{G}\mathbf{y}| \leqq |\mathbf{G}| \ |\mathbf{y}| \ .$$

Also, if $\mathbf{G}$ and $\mathbf{H}$ have the same dimension, we readily verify that

$$|\mathbf{G}+\mathbf{H}| \leqq |\mathbf{G}| + |\mathbf{H}| \tag{9}$$

and

$$|\mathbf{G}\mathbf{H}| \leqq |\mathbf{G}| \ |\mathbf{H}| \ . \tag{10}$$

A constructive method for finding $|\mathbf{G}|$ is to consider the constrained maximization problem

$$L(\mathbf{x}) = |\mathbf{G}\mathbf{x}|^2 - \lambda|\mathbf{x}|^2$$

where $|\mathbf{x}| = 1$ and λ is a Lagrange multiplier. Then setting

$$\frac{d}{d\mathbf{x}} L(\mathbf{x}) = 2\mathbf{G}'\mathbf{G}\mathbf{x} - 2\lambda\mathbf{x} \tag{11}$$

equal to the zero vector implies that $\mathbf{x}$ must be an eigenvector of $\mathbf{G}'\mathbf{G}$ and have norm one.

Now if $\mathbf{G}$ is any square matrix, $\mathbf{G}'\mathbf{G}$ is nonnegative definite since $\mathbf{x}'(\mathbf{G}'\mathbf{G})\mathbf{x} = |\mathbf{G}\mathbf{x}|^2 \geqq 0$ for all vectors $\mathbf{x}$. Thus all the eigenvalues of $\mathbf{G}'\mathbf{G}$ are nonnegative. If $\partial\mathbf{G} = n$, let λ_i with

$$\lambda_1 \geqq \lambda_2 \geqq \cdots \geqq \lambda_n \geqq 0$$

be the eigenvalues of $\mathbf{G}'\mathbf{G}$; and let $\mathbf{e}_1, \mathbf{e}_2, \cdots, \mathbf{e}_n$ be the corresponding eigenvectors of norm one. Then from (11)

$$\mathbf{G}'\mathbf{G}\mathbf{e}_m = \lambda_m\mathbf{e}_m \ , \ 1 \leqq m \leqq n$$

and

$$|\mathbf{G}\mathbf{e}_m|^2 = \mathbf{e}_m'\mathbf{G}'\mathbf{G}\mathbf{e}_m = \lambda_m|\mathbf{e}_m|^2 = \lambda_m \ .$$

Thus

$$|\mathbf{G}| = \lambda_1^{\frac{1}{2}} \ . \tag{12}$$

If $\mathbf{G}$ is itself nonnegative definite and μ_1 its largest eigenvalue, then $\mu_1^2 = \lambda_1$ and

$$|\mathbf{G}| = \mu_1 \ . \tag{13}$$

Now suppose $\mathbf{G}$ is positive definite. Let λ_i with

$$\lambda_1 \geqq \lambda_2 \geqq \cdots \geqq \lambda_n > 0$$

be the eigenvalues of $\mathbf{G}$. If $\mathbf{Q}$ is an orthogonal matrix with the property that $\mathbf{Q}'\mathbf{G}\mathbf{Q} = \mathbf{D}$ where $\mathbf{D}$ is the diagonal matrix $\|\lambda_i \delta_{ij}\|$, then

$$\mathbf{x}'\mathbf{G}\mathbf{x} = \mathbf{y}'\mathbf{D}\mathbf{y}$$

where $\mathbf{y} = \mathbf{Q}'\mathbf{x}$. Then $|\mathbf{x}| = |\mathbf{y}|$,

$$\max_{|\mathbf{x}|=1} \mathbf{x}'\mathbf{G}\mathbf{x} = \lambda_1 = |\mathbf{G}| \tag{14}$$

and

$$\min_{|\mathbf{x}|=1} \mathbf{x}'\mathbf{G}\mathbf{x} = \lambda_n \ . \tag{15}$$

These results lead to the following theorem,

Theorem 1. Let $\mathbf{A} = \|a_{ij}\|$ be an arbitrary $n \times n$ matrix and let $\mathbf{G}$ be a positive definite matrix. Then $\mathbf{G} + \alpha\mathbf{A}$ is positive definite for all α with

$$|\alpha| < \frac{\lambda}{an} \tag{16}$$

where λ is the smallest eigenvalue of $\mathbf{G}$ and $a = \max_{i,j} |a_{ij}|$.

Proof. Let $\mathbf{x}$ be any vector of norm one. Then

$$|\mathbf{x}'\mathbf{A}\mathbf{x}| \leqq an$$

where $a = \max_{i,j} |a_{ij}|$. Thus by (15)

$$\mathbf{x}'(\mathbf{G} + \alpha\mathbf{A})\mathbf{x} \geqq \lambda - |\alpha| \ |\mathbf{x}'\mathbf{A}\mathbf{x}|$$

where $\lambda > 0$ is the smallest eigenvalue of $\mathbf{G}$. The above expression is positive if α satisfies the inequality of (16). ■

While we're on the subject of differentiation, it is perhaps not inappropriate to consider the problem of differentiating a determinant with respect to a parameter. Of course this is ''just'' differentiation of a scalar with respect to a scalar. Nevertheless, we shall obtain some very useful formulas that will have future applications.

Suppose, then, that $\mathbf{G}$ is a square matrix which is a function of a real variable x. Let $\mathbf{G}=\mathbf{G}(x)$ be a continuously differentiable function of x for all x in some interval I. Then by definition of the derivative of a determinant

$$\frac{d}{dx}\det \mathbf{G}(x)=\sum_{i,j} G_{ij}\frac{d}{dx}\, g_{ij} \tag{17}$$

where $\mathbf{G}(x)=\|g_{ij}\|$ and G_{ij} is the cofactor of g_{ij}. Thus

$$\frac{d}{dx}\det \mathbf{G}(x)=\operatorname{tr}\mathcal{G}\,\frac{d}{dx}\,\mathbf{G} \tag{18}$$

where $\mathcal{G}'$ is the matrix of the cofactors G_{ij}.

We also note that since

$$\det \mathbf{G}(x)=\sum_{j} g_{ij}G_{ij}\ , \ \text{any } i$$

we may write

$$\begin{aligned}\det \mathbf{G}(x)&=\frac{1}{n}\sum_{i,j} g_{ij}G_{ij}\\ &=\frac{1}{n}\operatorname{tr}\mathbf{G}\mathcal{G}\end{aligned} \tag{19}$$

where $\partial\mathbf{G}=n$. If we differentiate (19) with respect to x and compare with (18), we have the further identity

$$(n-1)\operatorname{tr}\mathcal{G}\,\frac{d}{dx}\,\mathbf{G}=\operatorname{tr}\mathbf{G}\,\frac{d\mathcal{G}}{dx}\ . \tag{20}$$

The most useful version of the above result occurs when $\mathbf{G}(x)$ is nonsingular, and is expressed in the theorem below.

Theorem 2. Let $\mathbf{G}$ be a square matrix which is a continuously differentiable function of the real variable x for all x in an interval I. If $\det \mathbf{G}(x)$ is bounded away from zero on I, then

$$\frac{d}{dx}\det \mathbf{G}(x)=(\det \mathbf{G})\operatorname{tr}\mathbf{G}^{-1}\frac{d}{dx}\,\mathbf{G}\ . \tag{21}$$

Proof. Let $\mathcal{G}'$ be the matrix of the cofactors of $\mathbf{G}$. Since $\det \mathbf{G}(x)\neq 0$

on I by hypothesis, $\mathbf{G}$ is nonsingular and we may write

$$\mathcal{G} = (\det \mathbf{G})\mathbf{G}^{-1} \ .$$

Now substitute this result in (18) to obtain (21). ■

Corollary 1. If $\mathbf{G}(x) = \mathbf{H} + x\mathbf{K}$ where $\mathbf{H}$ and $\mathbf{K}$ are independent of x, then

$$\frac{d}{dx} \det (\mathbf{H} + x\mathbf{K}) = \det (\mathbf{H} + x\mathbf{K}) \operatorname{tr} \mathbf{K}(\mathbf{H} + x\mathbf{K})^{-1} \ .$$

6. MATRIX INVERSION II—Inversion of partitioned matrices

Suppose $\mathbf{R}$ is a 2×2 nonsingular matrix,

$$\mathbf{R} = \begin{bmatrix} a & b \\ b^* & c \end{bmatrix} .$$

Then its inverse, say

$$\mathbf{R}^{-1} = \begin{bmatrix} u & v \\ v^* & w \end{bmatrix}$$

exists. Now

$$\mathbf{R}\mathbf{R}^{-1} = \begin{bmatrix} au + bv^* & av + bw \\ b^*u + cv^* & b^*v + cw \end{bmatrix} . \tag{1}$$

But $\mathbf{R}\mathbf{R}^{-1}$ is also equal to the 2×2 identity matrix,

$$\mathbf{R}\mathbf{R}^{-1} = \begin{bmatrix} 1 & 0 \\ 0 & 1 \end{bmatrix} . \tag{2}$$

One way to find $\mathbf{R}^{-1}$ (that is, to determine u, v, v^*, w in terms of a, b, b^*, c) is to equate (1) and (2).

Let us apply this technique to the square partitioned matrix

$$\mathbf{R} = \begin{bmatrix} \mathbf{A} & \mathbf{B} \\ \mathbf{B}^* & \mathbf{C} \end{bmatrix} \tag{3}$$

(where $\mathbf{R}$, $\mathbf{A}$, $\mathbf{C}$ are assumed to be nonsingular) in an attempt to find $\mathbf{R}^{-1}$. If we write

$$\mathbf{R}^{-1} = \begin{bmatrix} \mathbf{U} & \mathbf{V} \\ \mathbf{V}^* & \mathbf{W} \end{bmatrix} , \tag{4}$$

then $\mathbf{R}\mathbf{R}^{-1} = \mathbf{I}$ implies

$$\mathbf{U} = (\mathbf{A} - \mathbf{B}\mathbf{C}^{-1}\mathbf{B}^*)^{-1} \tag{5}$$

$$\mathbf{W} = (\mathbf{C} - \mathbf{B}^*\mathbf{A}^{-1}\mathbf{B})^{-1} \tag{6}$$

$$\mathbf{V} = -\mathbf{A}^{-1}\mathbf{B}(\mathbf{C} - \mathbf{B}^*\mathbf{A}^{-1}\mathbf{B})^{-1} \tag{7}$$

$$\mathbf{V}^* = -\mathbf{C}^{-1}\mathbf{B}^*(\mathbf{A} - \mathbf{B}\mathbf{C}^{-1}\mathbf{B}^*)^{-1} . \tag{8}$$

Thus the problem of inverting $\mathbf{R}$ reduces to the problem of inverting lower order matrices.

Now $\mathbf{R}^{-1}\mathbf{R}$ also is the identity matrix, and $\mathbf{R}^{-1}\mathbf{R}=\mathbf{I}$ implies

$$\mathbf{V}=-(\mathbf{A}-\mathbf{B}\mathbf{C}^{-1}\mathbf{B}^*)^{-1}\mathbf{B}\mathbf{C}^{-1} \tag{9}$$

and

$$\mathbf{V}^*=-(\mathbf{C}-\mathbf{B}^*\mathbf{A}^{-1}\mathbf{B})^{-1}\mathbf{B}^*\mathbf{A}^{-1}\ . \tag{10}$$

So (9) and (10) give different formulas for $\mathbf{V}$ and $\mathbf{V}^*$ than do (7) and (8). The equality of (7) and (9) as well as the equality of (8) and (10) follows from the uniqueness of matrix inversion. (A direct proof may be given using Lemma 1 of Section 3.)

A special case of great usefulness arises when $\mathbf{C}=c$ is a nonzero scalar, $\mathbf{B}=\mathbf{b}$ is a column vector, and $\mathbf{B}^*=\boldsymbol{\beta}'$ is a row vector. Then

$$\mathbf{R}=\begin{bmatrix}\mathbf{A} & \mathbf{b}\\ \boldsymbol{\beta}' & c\end{bmatrix}\ . \tag{11}$$

Applying (5) – (8) in this case leads to

$$\mathbf{R}^{-1}=\begin{bmatrix}\left(\mathbf{A}-\frac{1}{c}\,\mathbf{b}\boldsymbol{\beta}'\right)^{-1} & -\frac{1}{c-\gamma}\mathbf{A}^{-1}\mathbf{b}\\[2ex] -\frac{1}{c-\gamma}\,\boldsymbol{\beta}'\mathbf{A}^{-1} & \frac{1}{c-\gamma}\end{bmatrix} \tag{12}$$

where

$$\gamma=\boldsymbol{\beta}'\mathbf{A}^{-1}\mathbf{b}\ .$$

If $\mathbf{b}=\mathbf{0}$, we may write (12) as

$$\mathbf{R}^{-1}=\begin{bmatrix}\mathbf{A}^{-1} & \mathbf{0}\\[1ex] -\frac{1}{c}\,\boldsymbol{\beta}'\mathbf{A}^{-1} & \frac{1}{c}\end{bmatrix} \tag{13}$$

with a similar formula if $\boldsymbol{\beta}=\mathbf{0}$. Thus we may express $\mathbf{R}^{-1}$ in terms of $\mathbf{A}^{-1}$. But even if both $\mathbf{b}$ and $\boldsymbol{\beta}$ are nonzero we may do this. For in this case $\mathbf{b}\boldsymbol{\beta}'$ has rank one, and by the Sherman-Morrison formula (3.8) (or Corollary 4.1)

$$\left(\mathbf{A}-\frac{1}{c}\,\mathbf{b}\boldsymbol{\beta}'\right)^{-1}=\mathbf{A}^{-1}+\frac{1}{c-\gamma}\,\mathbf{A}^{-1}\mathbf{b}\boldsymbol{\beta}'\mathbf{A}^{-1}\ .$$

Thus

$$\mathbf{R}^{-1} = \frac{1}{c-\gamma}\begin{bmatrix} (c-\gamma)\mathbf{A}^{-1} + \mathbf{A}^{-1}\mathbf{b}\boldsymbol{\beta}'\mathbf{A}^{-1} & -\mathbf{A}^{-1}\mathbf{b} \\ -\boldsymbol{\beta}'\mathbf{A}^{-1} & 1 \end{bmatrix} . \tag{14}$$

Hence, if $\mathbf{A}^{-1}$ is known, we have calculated, explicitly, the inverse of $\mathbf{R}$.

One also may make a direct assault on (11) by writing

$$\mathbf{R} = \mathbf{G} + \mathbf{H} \tag{15}$$

where

$$\mathbf{G} = \begin{bmatrix} \mathbf{A} & \mathbf{b} \\ \mathbf{0}' & c \end{bmatrix} .$$

and

$$\mathbf{H} = \begin{bmatrix} \mathbf{0} & \mathbf{0} \\ \boldsymbol{\beta}' & 0 \end{bmatrix} .$$

If $\boldsymbol{\beta} = \mathbf{0}$,

$$\mathbf{R}^{-1} = \mathbf{G}^{-1} = \begin{bmatrix} \mathbf{A}^{-1} & -\frac{1}{c}\mathbf{A}^{-1}\mathbf{b} \\ \mathbf{0}' & \frac{1}{c} \end{bmatrix} . \tag{16}$$

[see (13)] and we have found $\mathbf{R}^{-1}$. If $\boldsymbol{\beta} \neq \mathbf{0}$, then $\mathbf{H}$ has rank one and we may apply Corollary 4.1 to (15). [Note that (13) and (16) are special cases of (14) with $\mathbf{b} = \mathbf{0}$ or $\boldsymbol{\beta} = \mathbf{0}$ respectively.]

In terms of the decomposition of (15),

$$\gamma = \boldsymbol{\beta}'\mathbf{A}^{-1}\mathbf{b} = -c \operatorname{tr} \mathbf{H}\mathbf{G}^{-1} . \tag{17}$$

We also see from (15) that

$$\begin{aligned} \det \mathbf{R} = \det (\mathbf{G} + \mathbf{H}) &= \det \mathbf{G}(\mathbf{I} + \mathbf{G}^{-1}\mathbf{H}) \\ &= (\det \mathbf{G})(1 + \operatorname{tr} \mathbf{G}^{-1}\mathbf{H}) \\ &= (c-\gamma)(\det \mathbf{A}) . \end{aligned} \tag{18}$$

7. POSITIVE DEFINITE MATRICES

Positive definite matrices play a distinguished role in many matrix problems. We have seen a few of these applications in earlier sections. In particular they figure prominently in probabilistic theory since the covariance matrix is always positive definite (or nonnegative definite), see Appendix A. Let us therefore first recall some more or less standard results about such matrices; and then we shall consider some not so standard formulas involving positive definite matrices.

Suppose, then, that $\mathbf{G}$ and $\mathbf{F}$ are square matrices of the same dimension. Further, we shall suppose that $\mathbf{G}$ is positive definite and that $\mathbf{F}$ is nonnegative definite. Then there exists a nonsingular matrix $\mathbf{S}$ such that

$$\mathbf{S}'\mathbf{G}\mathbf{S} = \mathbf{I} \ .$$

Since $\mathbf{S}'\mathbf{F}\mathbf{S}$ is symmetric, there exists an orthogonal matrix $\mathbf{\Theta}$ such that

$$\mathbf{\Theta}'(\mathbf{S}'\mathbf{F}\mathbf{S})\mathbf{\Theta} = \mathbf{D}$$

where $\mathbf{D}$ is a nonnegative definite diagonal matrix. Now let $\mathbf{Q} = \mathbf{S}\mathbf{\Theta}$. Then

$$\mathbf{Q}'\mathbf{G}\mathbf{Q} = \mathbf{I} \tag{1}$$

and

$$\mathbf{Q}'\mathbf{F}\mathbf{Q} = \mathbf{D} \tag{2}$$

and we have *simultaneously diagonalized* $\mathbf{G}$ and $\mathbf{F}$.

Furthermore, since

$$\det(\mathbf{D} - \nu\mathbf{I}) = (\det \mathbf{Q})^2 \det(\mathbf{F} - \nu\mathbf{G})$$

we see that the eigenvalues ν_i of $\mathbf{D}$ are identical with the roots of the characteristic equation $\det(\mathbf{F} - \nu\mathbf{G}) = 0$.

Now suppose $\mathbf{G} - \mathbf{F}$ is nonnegative definite. Then

$$\mathbf{Q}'(\mathbf{G} - \mathbf{F})\mathbf{Q} = \mathbf{I} - \mathbf{D}$$

is nonnegative definite. But $\mathbf{D}$ also is nonnegative definite. Thus $0 \leqq \nu_i \leqq 1$ for all i. Since

$$\det \mathbf{I} = (\det \mathbf{Q})^2(\det \mathbf{G}) = 1$$

and

$$\det \mathbf{D} = (\det \mathbf{Q})^2(\det \mathbf{F}) = \prod_i \nu_i \leqq 1$$

we see that

$$\det \mathbf{G} \geqq \det \mathbf{F} \tag{3}$$

whenever $\mathbf{G}-\mathbf{F}$ is nonnegative definite.

Let us apply these formulas to a result of the previous section. Suppose $\mathbf{R}$ of (6.3) is positive definite. Then $\mathbf{A}$, $\mathbf{C}$, $\mathbf{U}$, $\mathbf{W}$ as well as their inverses are positive definite. Since $\mathbf{W}^{-1}$ is positive definite, (6.6) implies that

$$\mathbf{C}-\mathbf{B}'\mathbf{A}^{-1}\mathbf{B}$$

also is positive definite. (Note that the positive definite character of $\mathbf{R}$ implies that $\mathbf{B}^* = \mathbf{B}'$.) Hence

$$\det \mathbf{C} \geqq \det \mathbf{B}'\mathbf{A}^{-1}\mathbf{B}$$

follows from (3). In particular, if $\mathbf{B}$ is a square matrix,

$$(\det \mathbf{A})(\det \mathbf{C}) \geqq (\det \mathbf{B})^2 \ . \tag{4}$$

In Lemma 3.2 we proved that certain combinations of matrices were positive definite. We wish to consider some additional theorems in this vein.

Lemma 1. Let $\mathbf{A}$ be an $r \times n$ matrix of rank r. Then

$$\mathbf{S}=\mathbf{I}-\mathbf{A}'(\mathbf{A}\mathbf{A}')^{-1}\mathbf{A}$$

is nonnegative definite and has rank $n-r$. Furthermore, $n-r$ of the eigenvalues of $\mathbf{S}$ are $+1$ and the remaining r eigenvalues are zero.

Proof. Since $\mathbf{A}\mathbf{A}'$ is positive definite

$$\mathbf{T}=\mathbf{A}'(\mathbf{A}\mathbf{A}')^{-1}\mathbf{A}$$

is nonnegative definite. Hence all the eigenvalues of $\mathbf{T}$ are nonnegative. But $\mathbf{T}=\mathbf{T}^m$ for any positive integer m. Thus the eigenvalues of $\mathbf{T}$ must all be zeros and ones. Since $\rho\mathbf{T}=r$, the matrix $\mathbf{T}$ has r eigenvalues equal to $+1$ and $n-r$ eigenvalues equal to zero. Hence $\mathbf{S}=\mathbf{I}-\mathbf{T}$ is nonnegative definite with rank $n-r$. ■

From this result we see that

Theorem 1. Let $\mathbf{R}$ be an $n \times n$ positive definite matrix and let $\mathbf{B}$ be an

$r \times n$ matrix of rank r. Then

$$\mathbf{R}^{-1} - \mathbf{B}'(\mathbf{BRB}')^{-1}\mathbf{B}$$

is nonnegative definite with rank $n-r$.

Proof. Let $\mathbf{R} = \mathbf{PP}'$ and define $\mathbf{A}$ as $\mathbf{BP}$. Then $\mathbf{A}$ has rank r and by Lemma 1,

$$\begin{aligned}\mathbf{S} &= \mathbf{I} - \mathbf{A}'(\mathbf{AA}')^{-1}\mathbf{A} \\ &= \mathbf{I} - \mathbf{P}'\mathbf{B}'(\mathbf{BRB}')^{-1}\mathbf{BP}\end{aligned}$$

is nonnegative definite with rank $n-r$. Since $\mathbf{P}'^{-1}\mathbf{SP}^{-1}$ is also nonnegative definite with rank $n-r$, our theorem is proved. ■

Corollary 1. Under the hypotheses of Theorem 1,

$$(\det \mathbf{BRB}')(\det \mathbf{BR}^{-1}\mathbf{B}') \geqq (\det \mathbf{BB}')^2 \ .$$

Proof. From Theorem 1

$$\mathbf{BR}^{-1}\mathbf{B}' - (\mathbf{BB}')(\mathbf{BRB}')^{-1}(\mathbf{BB}')$$

is nonnegative definite. Thus

$$\det \mathbf{BR}^{-1}\mathbf{B}' \geqq \det (\mathbf{BB}')(\mathbf{BRB}')^{-1}(\mathbf{BB}')$$

by (3). ■

If $r = 1$ in Theorem 1, then $\mathbf{B}$ is an n-dimensional row vector, say $\mathbf{x}'$. Thus Corollary 1 implies that

$$(\mathbf{x}'\mathbf{Rx})(\mathbf{x}'\mathbf{R}^{-1}\mathbf{x}) \geqq |\mathbf{x}|^4 \ . \tag{5}$$

Actually we can prove a more general version of (5).

Theorem 2. Let $\mathbf{R}$ be a positive definite matrix. Then

$$(\mathbf{x}'\mathbf{Rx})(\mathbf{y}'\mathbf{R}^{-1}\mathbf{y}) \geqq (\mathbf{x}'\mathbf{y})^2$$

where $\mathbf{x}$ and $\mathbf{y}$ are vectors of dimension $\partial\mathbf{R}$.

Proof. Let $\mathbf{R} = \mathbf{PP}'$. Then for any real scalar β

$$|\mathbf{P}^{-1}\mathbf{y} + \beta\mathbf{P}'\mathbf{x}| \geqq 0 \ .$$

If $\mathbf{x}=\mathbf{0}$, the theorem is trivial. Now assume $\mathbf{x} \neq \mathbf{0}$ and let

$$\beta = -\frac{\mathbf{x}'\mathbf{y}}{\mathbf{x}'\mathbf{R}\mathbf{x}} \ . \ \blacksquare$$

Returning to (5) let us suppose that $\mathbf{R}$ is a nonsingular diagonal matrix, say $\mathbf{R}=\|\lambda_i\delta_{ij}\|$. Then $\mathbf{R}^{-1}=\|\lambda_i^{-1}\delta_{ij}\|$ and (5) yields the interesting inequality

$$\left(\sum_{i=1}^{n} \lambda_i x_i^2\right)\left(\sum_{i=1}^{n} \lambda_i^{-1} x_i^2\right) \geqq |\mathbf{x}|^4 \ . \tag{6}$$

Now using the *Kantorovich inequality* we may obtain an *upper* bound for the above inequality.

The Kantorovich inequality may be formulated as follows: Let $\lambda_1, \cdots, \lambda_n$ be a set of positive numbers and let $\mu_1, \cdots, \mu_n$ be a set of nonnegative numbers. Let $\boldsymbol{\lambda}'=\{\lambda_1, \cdots, \lambda_n\}$, $\boldsymbol{\Lambda}'=\{\lambda_1^{-1}, \cdots, \lambda_n^{-1}\}$, $\boldsymbol{\mu}'=\{\mu_1, \cdots, \mu_n\}$. Then

$$\boldsymbol{\mu}'\boldsymbol{\lambda}\boldsymbol{\Lambda}'\boldsymbol{\mu} \leqq \frac{\sigma^2(\alpha+\beta)^2}{4\alpha\beta} \tag{7}$$

where $\alpha=\max_i \lambda_i$, $\beta=\min_i \lambda_i$ and $\sigma=\sum_{i=1}^n \mu_i$.

We sketch a proof of this formula. Multiply the inequality

$$\frac{1}{\lambda_i} \leqq \frac{\alpha+\beta-\lambda_i}{\alpha\beta}$$

(which is valid for all i) by μ_i and sum over i to obtain

$$(\alpha+\beta)\sigma - \boldsymbol{\mu}'\boldsymbol{\lambda} \geqq \alpha\beta\boldsymbol{\mu}'\boldsymbol{\Lambda} \ . \tag{8}$$

Now for any real number k,

$$4(\alpha+\beta-k)k \leqq (\alpha+\beta)^2 \ . \tag{9}$$

If we let $k=\dfrac{1}{\sigma}\boldsymbol{\mu}'\boldsymbol{\lambda}$, then (8) and (9) yield (7).

Combining (6) and (7) we have:

Theorem 3. Let $\lambda_1, \cdots, \lambda_n$ be n positive numbers and let $\mathbf{x}'=$

$\{x_1, \cdots, x_n\}$. Then

$$|\mathbf{x}|^4 \leqq \left(\sum_{i=1}^{n} \lambda_i x_i^2\right)\left(\sum_{i=1}^{n} \lambda_i^{-1} x_i^2\right) \leqq \frac{1}{4}\left(\frac{\alpha}{\beta}+\frac{\beta}{\alpha}+2\right)|\mathbf{x}|^4$$

where $\alpha = \max_i \lambda_i$ and $\beta = \min_i \lambda_i$.

We mentioned at the beginning of this section that a covariance matrix is positive definite (or nonnegative definite). The matrix

$$\mathbf{R} = \|r_{ij}\| \ , \ \partial\mathbf{R} = n$$

where

$$\mathrm{r}_{ij} = \min\,(i,j)$$

occurs in the theory of Brownian motion (see Appendix A). It is asserted that $\mathbf{R}$ is positive definite. Let us prove this contention.

Let $\mathbf{x}' = \{x_1, \cdots, x_n\}$ be an arbitrary n-dimensional vector and q the quadratic form $\mathbf{x}'\mathbf{R}\mathbf{x}$. We shall show that $q \geqq 0$ for all $\mathbf{x}$, and that $q=0$ if and only if $\mathbf{x}=\mathbf{0}$. This will establish the positive definite character of $\mathbf{R}$. To do this we write

$$\begin{aligned} q = \mathbf{x}'\mathbf{R}\mathbf{x} &= \sum_{i=1}^{n}\sum_{j=1}^{n} r_{ij} x_i x_j \\ &= \sum_{i=1}^{n}\sum_{j=1}^{n} \min\,(i,j)\, x_i x_j \\ &= \sum_{i=1}^{n}\sum_{j=1}^{i} j x_i x_j + \sum_{i=1}^{n}\sum_{j=i+1}^{n} i x_i x_j \\ &= \sum_{j=1}^{n} j x_j \sum_{i=j}^{n} x_i + \sum_{j=1}^{n} j x_j \sum_{i=j+1}^{n} x_i \end{aligned} \tag{10}$$

where we have interchanged the order of summation in the first term and interchanged i and j in the second.

Now let

$$y_k = x_k + x_{k+1} + \cdots + x_n \ , \ 1 \leqq k \leqq n$$

$$y_{n+1} = 0 \ .$$

Then

$$y_k^2 - y_{k+1}^2 = x_k(x_k + 2x_{k+1} + \cdots + 2x_n) \ , \ 1 \leqq k \leqq n$$

and

$$\sum_{k=1}^{n} y_k = \sum_{k=1}^{n} kx_k \ . \tag{11}$$

Hence

$$q = \sum_{j=1}^{n} j(y_j^2 - y_{j+1}^2) = \sum_{k=1}^{n} y_k^2 \ . \tag{12}$$

Therefore $q \geqq 0$. Furthermore, $q=0$ if and only if $y_k=0$ for $k=1, \cdots, n$. But $x_k = y_k - y_{k+1}$, $1 \leqq k \leqq n$. Thus the vanishing of the y_k implies that $\mathbf{x}=\mathbf{0}$.

The same argument may be applied to show that

$$\mathbf{S} = \|s_{ij}\| \ , \ \partial\mathbf{S} = n \tag{13}$$

where

$$s_{ij} = \begin{cases} i(n+1-j) \ , \ j \geqq i \\ \\ j(n+1-i) \ , \ i \geqq j \end{cases} , \ 1 \leqq i,j \leqq n$$

is positive definite. For if we write s_{ij} as

$$s_{ij} = \min(i,j)[(n+1) - \max(i,j)] \ ,$$

then for an arbitrary vector $\mathbf{x}$,

$$Q = \mathbf{x}'\mathbf{S}\mathbf{x} = (n+1)q - \sum_{i=1}^{n}\sum_{j=1}^{n} ijx_ix_j$$

where q is given by (10). But from (12) and (11)

$$Q = (n+1)\sum_{k=1}^{n} y_k^2 - \left(\sum_{k=1}^{n} y_k\right)^2 \ . \tag{14}$$

Thus slightly rearranging the terms in (14),

$$Q = \sum_{k=1}^{n} y_k^2 + \frac{1}{2}\sum_{j=1}^{n}\sum_{k=1}^{n}(y_j - y_k)^2$$

—which is nonnegative.

8. MATRICES OF RANK ONE

Matrices of rank one have many interesting properties, some of which have appeared in previous sections. For example, in the proof of the fundamental Theorem 1 of Section 4 (and elsewhere) we used the fact that

$$\det(\mathbf{I}+\mathbf{H}) = 1 + \mathrm{tr}\ \mathbf{H}$$

where $\rho\mathbf{H} = 1$. We also shall have occasion to use matrices of rank one in subsequent sections. Let us therefore consider a few additional results concerning such matrices.

Suppose, then, that $\mathbf{H}$ is an $n \times n$ dimensional square matrix of rank one. Then we observed in Section 4 that $n-1$ eigenvalues of $\mathbf{H}$ were zero and the remaining eigenvalue was tr $\mathbf{H}$. Also we noted that there existed nonzero vectors, say $\mathbf{u}$ and $\mathbf{v}$, such that $\mathbf{H} = \mathbf{u}\mathbf{v}'$. Using these two elementary facts we shall show that

$$|\mathbf{H}| = |\mathbf{u}|\ |\mathbf{v}| \tag{1}$$

where $|\mathbf{H}|$ is the norm of $\mathbf{H}$ (see Section 5). To prove this we write

$$\mathbf{H}'\mathbf{H} = \mathbf{v}\mathbf{u}'\mathbf{u}\mathbf{v}' = |\mathbf{u}|^2\mathbf{v}\mathbf{v}' \ .$$

Since $\mathbf{v}\mathbf{v}'$ has rank one, the only nonzero eigenvalue of $\mathbf{H}'\mathbf{H}$ is $|\mathbf{u}|^2 \cdot$ tr $\mathbf{v}\mathbf{v}' = |\mathbf{u}|^2\ |\mathbf{v}|^2$. Hence (1) follows from (5.12).

Since

$$\mathrm{tr}\ \mathbf{H} = \mathrm{tr}\ \mathbf{u}\mathbf{v}' = \mathbf{v}'\mathbf{u} \tag{2}$$

and by the Schwarz inequality $|\mathbf{v}'\mathbf{u}| \leqq |\mathbf{u}|\ |\mathbf{v}|$, we see from (1) and (2) that

$$|\mathrm{tr}\ \mathbf{H}| \leqq |\mathbf{H}| \ . \tag{3}$$

Let $\mathbf{G}$ be any $n \times n$ matrix. Then

$$\mathbf{HGH} = \mathbf{u}\mathbf{v}'\mathbf{G}\mathbf{u}\mathbf{v}' = \mathbf{u}(\mathbf{v}'\mathbf{G}\mathbf{u})\mathbf{v}' = (\mathbf{v}'\mathbf{G}\mathbf{u})\mathbf{H} \ . \tag{4}$$

But we may write the bilinear form $\mathbf{v}'\mathbf{G}\mathbf{u}$ as

$$\mathbf{v}'\mathbf{G}\mathbf{u} = \mathrm{tr}\ \mathbf{u}\mathbf{v}'\mathbf{G} = \mathrm{tr}\ \mathbf{HG} \ .$$

Hence from (4)

$$\mathbf{HGH} = (\mathrm{tr}\ \mathbf{HG})\mathbf{H} \ . \tag{5}$$

We call (5) the *reproducing property* of matrices of rank one. Note that no explicit mention of the vectors $\mathbf{u}$ and $\mathbf{v}$ appears in the formulation of

(5). Iteration of (5) leads to

$$(\mathbf{HG})^m = (\text{tr } \mathbf{HG})^{m-1}\mathbf{HG} \tag{6}$$

for all positive integers m.

If $\mathbf{F}$ is another $n \times n$ matrix, then from (5)

$$\mathbf{HGHF} = (\text{tr } \mathbf{HG})\mathbf{HF}$$

and

$$\text{tr } \mathbf{HGHF} = (\text{tr } \mathbf{HG})(\text{tr } \mathbf{HF}) \ . \tag{7}$$

In general, though, if $\mathbf{K}$ also has rank one,

$$\text{tr } \mathbf{HK} \neq (\text{tr } \mathbf{H})(\text{tr } \mathbf{K}) \ . \tag{8}$$

For, let $\mathbf{H} = \|h_{ij}\|$ and $\mathbf{K} = \|k_{ij}\|$. Then if $h_{11} = 1 = k_{nn}$ and all other h_{ij} and k_{ij} are zero, tr $\mathbf{H} = 1 =$ tr $\mathbf{K}$, but tr $\mathbf{HK} = 0$.

Suppose now that $\mathbf{P}$ is a square matrix of norm less than one. Then $(\mathbf{I}-\mathbf{P})^{-1}$ has the *Neumann series* expansion

$$(\mathbf{I}-\mathbf{P})^{-1} = \mathbf{I} + \mathbf{P} + \mathbf{P}^2 + \cdots + \mathbf{P}^m + \cdots \tag{9}$$

(see [Halmos, p. 186]). If $\rho\mathbf{P} = 1$, then by (6) we have $\mathbf{P}^m = (\text{tr } \mathbf{P})^{m-1}\mathbf{P}$ and we may write (9) as

$$(\mathbf{I}-\mathbf{P})^{-1} = \mathbf{I} + [1 + (\text{tr } \mathbf{P}) + \cdots + (\text{tr } \mathbf{P})^{m-1} + \cdots]\mathbf{P} \ .$$

But $|\text{tr } \mathbf{P}| \leqq |\mathbf{P}|$ by (3), and $|\mathbf{P}| < 1$ by assumption. Hence the above formula may be written as

$$(\mathbf{I}-\mathbf{P})^{-1} = \mathbf{I} + \frac{1}{1 - \text{tr } \mathbf{P}}\mathbf{P} \ .$$

This equation is just Theorem 4.1, and may be considered to be an "analytic continuation" of (9) for matrices of rank one.

Now let $\mathbf{A} = \|a_{ij}\|$ be an $n \times n$ nonsingular matrix and let $\|b_{ij}\|$ be its inverse. Then trivially

$$\sum_{k=1}^{n} a_{ik}b_{kj} = \delta_{ij} \ . \tag{10}$$

In particular, if the first row of $\mathbf{A}$ consists entirely of ones, that is $a_{11} = a_{12} = \cdots = a_{1n} = 1$, then from (10)

$$\sum_{k=1}^{n} b_{kj} = \delta_{1j}$$

and

$$\sum_{i,j=1}^{n} b_{ij}=1=\operatorname{tr} \mathbf{A}^{-1}\boldsymbol{\Delta} \tag{11}$$

where $\boldsymbol{\Delta}$ is an $n\times n$ matrix with ones everywhere. We also have

$$\begin{aligned} \operatorname{tr} \mathbf{A}\boldsymbol{\Delta}\mathbf{A}^{-1}\boldsymbol{\Delta} &= (\operatorname{tr} \mathbf{A}\boldsymbol{\Delta})(\operatorname{tr} \mathbf{A}^{-1}\boldsymbol{\Delta}) \\ &= \operatorname{tr} \mathbf{A}\boldsymbol{\Delta} = \sum_{i,j=1}^{n} a_{ij} \end{aligned} \tag{12}$$

by (7) and (11).

9. KRONECKER PRODUCTS

When we write a matrix $\mathbf{A}$ as $\mathbf{A}=\|a_{ij}\|$ it is tacitly assumed that the first subscript is the row index and the second subscript is the column index. Thus, for example, $\|a_{ij}\|=\|a_{ji}\|$. In this notation there is no way to represent the transpose $\mathbf{A}'$ of $\mathbf{A}$. For convenience in certain situations we shall write $\mathbf{A}=\|a_{ij}\|_{ij}$ where the outer set of subscripts is always in the order row, column, Thus $\|a_{ij}\|_{ij}=\|a_{ji}\|_{ji}$, but $\mathbf{A}'=\|a_{ij}\|_{ji}=\|a_{ji}\|_{ij}$. This notation and its extensions (to be introduced later) are useful in writing various complicated expressions in compact form, especially those involving partitioned matrices.

We have met partitioned matrices before, most recently in Section 6. For example,

$$\mathbf{R}=\begin{bmatrix} \mathbf{A}_1 & \mathbf{A}_2 \\ \mathbf{A}_3 & \mathbf{A}_4 \end{bmatrix} \tag{1}$$

is a partitioned matrix (where $\mathbf{A}_1$ and $\mathbf{A}_2$ have the same number of rows and $\mathbf{A}_1$ and $\mathbf{A}_3$ the same number of columns). Roughly speaking, (1) may be called a ''higher order'' matrix since $\mathbf{R}$ is a matrix whose entries are themselves matrices. Of course, if $\mathbf{A}_\alpha=\|a^{(\alpha)}_{ij}\|$, $\alpha=1,2,3,4$, then we could write $\mathbf{R}$ as one big matrix whose entries were the scalars $a^{(\alpha)}_{ij}$. However, one virtue of matrix notation is its compactness.

Now suppose $\mathbf{A}_\alpha=b_\alpha\mathbf{A}$, $\alpha=1,2,3,4$. That is, each $\mathbf{A}_\alpha$ is a scalar multiple of a common matrix $\mathbf{A}$. Then (1) may be written as

$$\mathbf{R}=\begin{bmatrix} b_1\mathbf{A} & b_2\mathbf{A} \\ b_3\mathbf{A} & b_4\mathbf{A} \end{bmatrix} . \tag{2}$$

This special form of a partitioned matrix occurs in a surprising number of respectable problems. Examining (2) we see that

$$\mathbf{B}=\begin{bmatrix} b_1 & b_2 \\ b_3 & b_4 \end{bmatrix}$$

may be considered to be a matrix. Is there any way to represent (2) in terms of $\mathbf{A}$ and $\mathbf{B}$? The answer is yes. We write

$$\mathbf{R}=\mathbf{A}\otimes\mathbf{B}$$

and call $\mathbf{R}$ the *Kronecker product* of $\mathbf{A}$ and $\mathbf{B}$.

More generally, let $\mathbf{A}=\|a_{ij}\|$ and $\mathbf{B}=\|b_{\alpha\beta}\|$. Then the Kronecker product of $\mathbf{A}$ and $\mathbf{B}$ is defined as the partitioned matrix

$$\mathbf{A} \otimes \mathbf{B} = \|b_{\alpha\beta}\mathbf{A}\| \tag{3}$$

(see [MacDuffee]). Thus if $\mathbf{A}$ is $m \times n$ dimensional and $\mathbf{B}$ is $p \times q$ dimensional, then $\mathbf{A} \otimes \mathbf{B}$ is $mp \times nq$ dimensional. In certain applications we find it convenient to write

$$\mathbf{A} \otimes \mathbf{B} = \|a_{ij}b_{\alpha\beta}\|_{ij,\alpha\beta} \ .$$

The outer set of subscripts $ij,\alpha\beta$ means that i,j is varied first, and then α,β. Thus, for example, we may write $\mathbf{B} \otimes \mathbf{A}$ as

$$\mathbf{B} \otimes \mathbf{A} = \|a_{ij}b_{\alpha\beta}\|_{\alpha\beta,ij} \ .$$

We may use the above notation to prove the associativity of the $\otimes$ operation.

Theorem 1. Let $\mathbf{A}$, $\mathbf{B}$, $\mathbf{C}$ be matrices. Then

$$(\mathbf{A} \otimes \mathbf{B}) \otimes \mathbf{C} = \mathbf{A} \otimes (\mathbf{B} \otimes \mathbf{C}) \ . \tag{4}$$

Proof. By definition

$$(\mathbf{A} \otimes \mathbf{B}) \otimes \mathbf{C} = \|a_{ij}b_{\alpha\beta}\|_{ij,\alpha\beta} \otimes \mathbf{C} = \|(a_{ij}b_{\alpha\beta})c_{\mu\nu}\|_{ij,\alpha\beta,\mu\nu}$$

where $\mathbf{C} = \|c_{\mu\nu}\|$. But

$$(a_{ij}b_{\alpha\beta})c_{\mu\nu} = a_{ij}(b_{\alpha\beta}c_{\mu\nu}) \ .$$

Thus

$$\begin{aligned}(\mathbf{A} \otimes \mathbf{B}) \otimes \mathbf{C} &= \|a_{ij}(b_{\alpha\beta}c_{\mu\nu})\|_{ij,\alpha\beta,\mu\nu} = \mathbf{A} \otimes \|b_{\alpha\beta}c_{\mu\nu}\|_{\alpha\beta,\mu\nu} \\ &= \mathbf{A} \otimes (\mathbf{B} \otimes \mathbf{C}) \ . \ \blacksquare\end{aligned}$$

We may generalize Theorem 1 by writing

$$\begin{aligned}\mathbf{A}_1 \otimes \mathbf{A}_2 \otimes \cdots \otimes \mathbf{A}_n &= \bigotimes_{\gamma=1}^{n} \mathbf{A}_\gamma \\ &= \|a^{(1)}_{i_1j_1}a^{(2)}_{i_2j_2} \cdots a^{(n)}_{i_nj_n}\|_{i_1j_1,i_2j_2,\cdots,i_nj_n}\end{aligned}$$

where $\mathbf{A}_\gamma = \|a^{(\gamma)}_{i_\gamma j_\gamma}\|$. Thus if $k_1, \cdots, k_n$ is a permutation of the integers $1, \cdots, n$,

$$\begin{aligned}\mathbf{A}_{k_1} \otimes \mathbf{A}_{k_2} \otimes \cdots \otimes \mathbf{A}_{k_n} &= \bigotimes_{\gamma=1}^{n} \mathbf{A}_{k_\gamma} \\ &= \|a^{(1)}_{i_1j_1}a^{(2)}_{i_2j_2} \cdots a^{(n)}_{i_nj_n}\|_{i_{k_1}j_{k_1},i_{k_2}j_{k_2},\cdots,i_{k_n}j_{k_n}} \ .\end{aligned}$$

One of the most useful formulas available for manipulating Kronecker products is given in Theorem 2.

Theorem 2. For compatible matrices **A**, **G** and **B**, **H**,

$$(\mathbf{A} \otimes \mathbf{B})(\mathbf{G} \otimes \mathbf{H}) = \mathbf{AG} \otimes \mathbf{BH} \ . \tag{5}$$

Proof. From (3)

$$(\mathbf{A} \otimes \mathbf{B})(\mathbf{G} \otimes \mathbf{H}) = \|b_{ij}\mathbf{A}\| \ \|h_{\alpha\beta}\mathbf{G}\| = \|\gamma_{j\beta}\mathbf{AG}\|$$

where $\mathbf{B} = \|b_{ij}\|$, $\mathbf{H} = \|h_{\alpha\beta}\|$ and

$$\gamma_{j\beta} = \sum_k b_{ik} h_{k\beta} \ .$$

But $\mathbf{BH} = \|\gamma_{j\beta}\|$. ■

The distributive law

$$\mathbf{A} \otimes (\mathbf{G} + \mathbf{H}) = \mathbf{A} \otimes \mathbf{G} + \mathbf{A} \otimes \mathbf{H} \tag{6a}$$

$$(\mathbf{G} + \mathbf{H}) \otimes \mathbf{A} = \mathbf{G} \otimes \mathbf{A} + \mathbf{H} \otimes \mathbf{A} \tag{6b}$$

also is valid for compatible matrices **G** and **H**. For,

$$\mathbf{A} \otimes (\mathbf{G} + \mathbf{H}) = \|(g_{ij} + h_{ij})\mathbf{A}\| = \|g_{ij}\mathbf{A}\| + \|h_{ij}\mathbf{A}\| \ ,$$

which, by (3), is (6a).

Trivially, the transpose is the product of the transposes,

$$(\mathbf{A} \otimes \mathbf{B})' = \mathbf{A}' \otimes \mathbf{B}' \ . \tag{7}$$

Note that the order of **A** and **B** is not reversed, as, for example, in the formula $(\mathbf{AB})' = \mathbf{B}'\mathbf{A}'$ (when the product of **A** and **B** is defined).

We have used (and shall continue to use) **I** to represent the identity matrix. The dimension of **I** must be inferred from the context. Frequently, especially when dealing with partitioned matrices, it is convenient to indicate the dimension of **I** by a subscript. Thus if **I** is the $n \times n$ identity matrix, we sometimes shall write **I** as $\mathbf{I}_n$ to indicate explicitly its dimension.

Now suppose **a** is an m-dimensional vector and **b** an n-dimensional vector. Then $\mathbf{a} \otimes \mathbf{b}$ is an mn-dimensional column vector and $\mathbf{a}' \otimes \mathbf{b}'$ is an mn-dimensional row vector. Sometimes it is convenient to write $\mathbf{a} \otimes \mathbf{b}$ in the form

$$\mathbf{a} \otimes \mathbf{b} = (\mathbf{a} \otimes \mathbf{I}_n)\mathbf{b}$$

and by (7)

$$\mathbf{a}'\otimes \mathbf{b}' = \mathbf{b}'(\mathbf{a}'\otimes \mathbf{I}_n) \ .$$

The Kronecker product of a row and column vector is a matrix, viz.:

$$\mathbf{a}' \otimes \mathbf{b} = \mathbf{b}\mathbf{a}' \tag{8}$$

and

$$\mathbf{a} \otimes \mathbf{b}' = \mathbf{a}\mathbf{b}' \ . \tag{9}$$

Finally we note that if λ is a scalar and $\mathbf{A}$ a matrix, then

$$\lambda \otimes \mathbf{A} = \mathbf{A} \otimes \lambda = \lambda\mathbf{A} \ . \tag{10}$$

The above results are the basic formulas for manipulating Kronecker products. In the next six theorems we shall show the relationship between the trace, inverse, rank, eigenvalues, norm, and determinant of a Kronecker product $\mathbf{A} \otimes \mathbf{B}$ in terms of the corresponding properties of $\mathbf{A}$ and $\mathbf{B}$.

Theorem 3. Let $\mathbf{A}$ and $\mathbf{B}$ be square matrices. Then

$$\text{tr}\ (\mathbf{A} \otimes \mathbf{B}) = (\text{tr}\ \mathbf{A})(\text{tr}\ \mathbf{B}) \ . \tag{11}$$

Proof. If we write $\mathbf{A} \otimes \mathbf{B} = \mathbf{A} \otimes \|b_{ij}\|$, then

$$\text{tr}\ (\mathbf{A} \otimes \mathbf{B}) = \sum_i \text{tr}\ \mathbf{A}b_{ii} = \left(\sum_i b_{ii}\right)(\text{tr}\ \mathbf{A}) \ . \quad \blacksquare$$

Theorem 4. If $\mathbf{A}$ and $\mathbf{B}$ are nonsingular matrices, then $\mathbf{A} \otimes \mathbf{B}$ is nonsingular and

$$(\mathbf{A} \otimes \mathbf{B})^{-1} = \mathbf{A}^{-1} \otimes \mathbf{B}^{-1} \ . \tag{12}$$

Proof. If $\partial\mathbf{A} = m$ and $\partial\mathbf{B} = n$, then Theorem 2 implies that

$$(\mathbf{A} \otimes \mathbf{B})(\mathbf{A}^{-1} \otimes \mathbf{B}^{-1}) = \mathbf{I}_m \otimes \mathbf{I}_n = \mathbf{I}_{mn} \ . \quad \blacksquare$$

Note that the order of $\mathbf{A}$ and $\mathbf{B}$ is not reversed, as, for example, in the formula $(\mathbf{AB})^{-1} = \mathbf{B}^{-1}\mathbf{A}^{-1}$ (when $m = n$).

In Theorem 5 below we shall show that $\rho(\mathbf{A} \otimes \mathbf{B}) = (\rho\mathbf{A})(\rho\mathbf{B})$. First, however, it is convenient to prove the following lemma.

Lemma 1. If $\mathbf{A}$ and $\mathbf{B}$ are nonsingular, then $\rho(\mathbf{A} \otimes \mathbf{B}) = (\rho\mathbf{A})(\rho\mathbf{B})$.

Proof. Since $\mathbf{A}$ and $\mathbf{B}$ are nonsingular, $\partial\mathbf{A} = \rho\mathbf{A}$ and $\partial\mathbf{B} = \rho\mathbf{B}$. By Theorem 4, $\partial(\mathbf{A} \otimes \mathbf{B}) = \rho(\mathbf{A} \otimes \mathbf{B})$. But by definition of the Kronecker product, $\partial(\mathbf{A} \otimes \mathbf{B}) = (\partial\mathbf{A})(\partial\mathbf{B})$. ■

Theorem 5. For any matrices $\mathbf{A}$ and $\mathbf{B}$

$$\rho(\mathbf{A} \otimes \mathbf{B}) = (\rho\mathbf{A})(\rho\mathbf{B}) \ . \tag{13}$$

Proof. Let $\rho\mathbf{A} = r$ and $\rho\mathbf{B} = s$. Choose nonsingular square matrices $\mathbf{P}$ and $\mathbf{Q}$ such that

$$\mathbf{PAQ} = \mathbf{A}^* = \|a^*_{ij}\|$$

where $a^*_{11} = a^*_{22} = \cdots = a^*_{rr} = 1$ and all other a^*_{ij} are zero. Similarly choose $\mathbf{R}$ and $\mathbf{S}$ such that $\mathbf{RBS} = \mathbf{B}^* = \|b^*_{ij}\|$ and $b^*_{11} = b^*_{22} = \cdots = b^*_{ss} = 1$, while all other b^*_{ij} are zero. Then since $\mathbf{P} \otimes \mathbf{R}$ and $\mathbf{Q} \otimes \mathbf{S}$ are nonsingular by Theorem 4,

$$\rho[(\mathbf{P} \otimes \mathbf{R})(\mathbf{A} \otimes \mathbf{B})(\mathbf{Q} \otimes \mathbf{S})] = \rho(\mathbf{A} \otimes \mathbf{B}) \ .$$

But $(\mathbf{P} \otimes \mathbf{R})(\mathbf{A} \otimes \mathbf{B})(\mathbf{Q} \otimes \mathbf{S}) = \mathbf{A}^* \otimes \mathbf{B}^*$ by (5). Clearly $\rho(\mathbf{A}^* \otimes \mathbf{B}^*) = rs$. ■

Theorem 6. Let $\partial\mathbf{A} = m$ with $\lambda_1, \cdots, \lambda_m$ the eigenvalues of $\mathbf{A}$, and let $\partial\mathbf{B} = n$ with $\mu_1, \cdots, \mu_n$ the eigenvalues of $\mathbf{B}$. Then the mn eigenvalues of $\mathbf{A} \otimes \mathbf{B}$ are

$$\lambda_i\mu_j \ , \ 1 \leqq i \leqq m \ , \ 1 \leqq j \leqq n \ . \tag{14}$$

Proof. Let $\mathbf{J}$ and $\mathbf{K}$ be the Jordan normal forms of $\mathbf{A}$ and $\mathbf{B}$ respectively. Then

$$\begin{aligned} \det(\mathbf{A} \otimes \mathbf{B} - \nu\mathbf{I} \otimes \mathbf{I}) &= \det(\mathbf{J} \otimes \mathbf{K} - \nu\mathbf{I} \otimes \mathbf{I}) \\ &= \prod_{j=1}^{n} \det(\mu_j\mathbf{J} - \nu\mathbf{I}) \\ &= \prod_{i=1}^{m} \prod_{j=1}^{n} (\mu_j\lambda_i - \nu) \ . \quad ■ \end{aligned}$$

This result can be used immediately to prove [see (8) of section 5]:

Theorem 7. Let $\mathbf{A}$ and $\mathbf{B}$ be square matrices. Then

$$|\mathbf{A} \otimes \mathbf{B}| = |\mathbf{A}|\ |\mathbf{B}| \ . \tag{15}$$

Proof. Let λ be the largest eigenvalue of $\mathbf{A}'\mathbf{A}$ and μ the largest eigenvalue of $\mathbf{B}'\mathbf{B}$. Now $(\mathbf{A} \otimes \mathbf{B})'(\mathbf{A} \otimes \mathbf{B}) = \mathbf{A}'\mathbf{A} \otimes \mathbf{B}'\mathbf{B}$ by (7) and (5); and by Theorem 6, the maximum eigenvalue of $\mathbf{A}'\mathbf{A} \otimes \mathbf{B}'\mathbf{B}$ is $\lambda\mu$. ■

As the last of this sequence of theorems we prove the result regarding determinants of Kronecker products.

Theorem 8. If $\partial\mathbf{A} = m$ and $\partial\mathbf{B} = n$, then

$$\det(\mathbf{A} \otimes \mathbf{B}) = (\det \mathbf{A})^n (\det \mathbf{B})^m \ . \tag{16}$$

Proof. Let $\mathbf{J}$ and $\mathbf{K}$ be the Jordan normal forms of $\mathbf{A}$ and $\mathbf{B}$ respectively. Then

$$\det(\mathbf{A} \otimes \mathbf{B}) = \det(\mathbf{J} \otimes \mathbf{K}) \ .$$

But $\mathbf{J} \otimes \mathbf{K}$ is a block upper triangular partitioned matrix with diagonal $[\mu_1\mathbf{J}, \mu_2\mathbf{J}, \cdots, \mu_n\mathbf{J}]$ where the μ_i are the eigenvalues of $\mathbf{B}$. Thus

$$\det(\mathbf{A} \otimes \mathbf{B}) = \left(\prod_{i=1}^{n} \mu_i^m\right)(\det \mathbf{J})^n$$

—which is (16). ■

Let us consider some applications of the previous theorem.

Theorem 9. Let $\partial\mathbf{A} = \partial\mathbf{B} = m$ and $\partial\mathbf{R} = \partial\mathbf{S} = n$. If $\mathbf{R}$ is positive definite and $\mathbf{S}$ nonnegative definite, then

$$\det(\mathbf{A} \otimes \mathbf{R} + \mathbf{B} \otimes \mathbf{S}) = (\det \mathbf{R})^m \prod_{i=1}^{n} \det(\mathbf{A} + \nu_i\mathbf{B}) \tag{17}$$

where the ν_i are the n roots of the characteristic equation $\det(\mathbf{S} - \nu\mathbf{R}) = 0$.

Proof. Let $\mathbf{P}$ be a nonsingular matrix such that

$$\mathbf{P}'\mathbf{R}\mathbf{P} = \mathbf{I}$$

and $\boldsymbol{\Theta}$ an orthogonal matrix such that

$$\boldsymbol{\Theta}'(\mathbf{P}'\mathbf{S}\mathbf{P})\boldsymbol{\Theta} = \|\nu_i\delta_{ij}\|$$

where the ν_i are the roots of the characteristic equation $\det(\mathbf{S}-\nu\mathbf{R})=0$, see Section 7.

By Theorem 2

$$(\mathbf{I}\otimes\mathbf{\Theta}')(\mathbf{I}\otimes\mathbf{P}')(\mathbf{A}\otimes\mathbf{R}+\mathbf{B}\otimes\mathbf{S})(\mathbf{I}\otimes\mathbf{P})(\mathbf{I}\otimes\mathbf{\Theta})=\mathbf{A}\otimes\mathbf{I}+\mathbf{B}\otimes\mathbf{D}$$

where $\mathbf{D}=\|\nu_i\delta_{ij}\|$. Thus

$$\begin{aligned}\det(\mathbf{A}\otimes\mathbf{I}+\mathbf{B}\otimes\mathbf{D})&=\prod_{i=1}^{n}\det(\mathbf{A}+\nu_i\mathbf{B})\\&=(\det\mathbf{P}'\mathbf{P})^m\det(\mathbf{A}\otimes\mathbf{R}+\mathbf{B}\otimes\mathbf{S})\ .\end{aligned}\tag{18}$$

But $\mathbf{PP}'=\mathbf{R}^{-1}$. ■

Now suppose we wish to find the determinant of $\mathbf{A}\otimes\mathbf{G}+\mathbf{B}\otimes\mathbf{E}$ where $\mathbf{G}$ and $\mathbf{E}$ do not necessarily have any definite properties. We can do this if $\mathbf{E}$ has rank one.

Theorem 10. Let $\partial\mathbf{A}=\partial\mathbf{B}=m$ and $\partial\mathbf{G}=\partial\mathbf{E}=n$ where $\mathbf{E}$ has rank one and $\mathbf{G}$ is nonsingular. Let

$$\mathbf{C}=\mathbf{A}\otimes\mathbf{G}+\mathbf{B}\otimes\mathbf{E}\ .$$

Then

$$\det\mathbf{C}=(\det\mathbf{G})^m(\det\mathbf{A})^{n-1}\det[\mathbf{A}+(\operatorname{tr}\mathbf{EG}^{-1})\mathbf{B}]\ .\tag{19}$$

Proof. We may write $\mathbf{C}$ as

$$\mathbf{C}=(\mathbf{A}\otimes\mathbf{I}+\mathbf{B}\otimes\mathbf{EG}^{-1})(\mathbf{I}\otimes\mathbf{G})$$

and

$$\det(\mathbf{A}\otimes\mathbf{I}+\mathbf{B}\otimes\mathbf{EG}^{-1})=\det(\mathbf{A}\otimes\mathbf{I}+\mathbf{B}\otimes\mathbf{J})$$

where $\mathbf{J}$ is the Jordan normal form of $\mathbf{EG}^{-1}$. As in Theorem 9 [see (18)]

$$\det(\mathbf{A}\otimes\mathbf{I}+\mathbf{B}\otimes\mathbf{J})=\prod_{i=1}^{n}\det(\mathbf{A}+\nu_i\mathbf{B})$$

where the ν_i are the eigenvalues of $\mathbf{EG}^{-1}$. Since $\rho\mathbf{EG}^{-1}=1$, all the ν_i are zero, except possibly $\operatorname{tr}\mathbf{EG}^{-1}$. ■

Corollary 1. Let $\mathbf{G}=\mathbf{I}$ in Theorem 10. Then if $\lambda_1,\cdots,\lambda_m$ are the eigenvalues of $\mathbf{A}$ and $\mu_1,\cdots,\mu_m$ the eigenvalues of $\mathbf{A}+(\operatorname{tr}\mathbf{E})\mathbf{B}$, then each

λ_i, $1 \leqq i \leqq m$, is an eigenvalue of multiplicity $n-1$ of $\mathbf{C}$ and the μ_i, $1 \leqq i \leqq m$, are simple eigenvalues of $\mathbf{C}$.

Proof. The characteristic polynomial of $\mathbf{C}$ is

$$\det(\mathbf{A} \otimes \mathbf{I} + \mathbf{B} \otimes \mathbf{E} - \lambda\mathbf{I} \otimes \mathbf{I}) = \det(\mathbf{A} - \lambda\mathbf{I})^{n-1}[\mathbf{A} + (\mathrm{tr}\,\mathbf{E})\mathbf{B} - \lambda\mathbf{I}] \ . \ \blacksquare$$

We now wish to develop some formulas analogous to the fundamental result of Corollary 4.1. Let

$$\mathbf{C} = \mathbf{A} \otimes \mathbf{B} + \mathbf{E} \otimes \mathbf{F} \tag{20}$$

where $\mathbf{A}$, $\mathbf{B}$, $\mathbf{E}$, $\mathbf{F}$ are square matrices with $\partial\mathbf{A} = \partial\mathbf{E}$, $\partial\mathbf{B} = \partial\mathbf{F}$ and $\rho\mathbf{F} = 1$. If we assume $\mathbf{C}$ to be nonsingular, and search for an inverse in the form

$$\mathbf{C}^{-1} = \mathbf{A}^{-1} \otimes \mathbf{B}^{-1} - \mathbf{Q} \otimes \mathbf{B}^{-1}\mathbf{F}\mathbf{B}^{-1} \ , \tag{21}$$

then multiplying (20) and (21) together implies that

$$\mathbf{Q} = [\mathbf{A} + (\mathrm{tr}\,\mathbf{F}\mathbf{B}^{-1})\mathbf{E}]^{-1}\mathbf{E}\mathbf{A}^{-1} \ . \tag{22}$$

If $\partial\mathbf{B} = \partial\mathbf{F} = 1$, then (21) reduces to

$$(\mathbf{A} + f\mathbf{E})^{-1} = \mathbf{A}^{-1} - f(\mathbf{A} + f\mathbf{E})^{-1}\mathbf{E}\mathbf{A}^{-1} \tag{23}$$

where $f = \mathrm{tr}\,\mathbf{F}\mathbf{B}^{-1}$. But (23) is just (2) of Lemma 3.1.

Now, returning to (21), suppose that $\mathbf{E}$ also has rank one. Then by Corollary 4.1 [see (22)]

$$[\mathbf{A} + (\mathrm{tr}\,\mathbf{F}\mathbf{B}^{-1})\mathbf{E}]^{-1} = \mathbf{A}^{-1} - \frac{\mathrm{tr}\,\mathbf{F}\mathbf{B}^{-1}}{1 + (\mathrm{tr}\,\mathbf{F}\mathbf{B}^{-1})(\mathrm{tr}\,\mathbf{E}\mathbf{A}^{-1})}\mathbf{A}^{-1}\mathbf{E}\mathbf{A}^{-1} \ .$$

Substituting this result in (21) leads to the elegant formula

$$(\mathbf{A} \otimes \mathbf{B} + \mathbf{E} \otimes \mathbf{F})^{-1} = \mathbf{A}^{-1} \otimes \mathbf{B}^{-1} - \frac{1}{1 + (\mathrm{tr}\,\mathbf{E}\mathbf{A}^{-1})(\mathrm{tr}\,\mathbf{F}\mathbf{B}^{-1})}\mathbf{A}^{-1}\mathbf{E}\mathbf{A}^{-1} \otimes \mathbf{B}^{-1}\mathbf{F}\mathbf{B}^{-1} \ . \tag{24}$$

10. ELEMENTARY MATRICES

Let $\mathbf{E}=\|e_{ij}\|$. If $e_{\mu\nu}=1$ for some μ and ν while $e_{ij}=0$ if $i \neq \mu$ or $j \neq \nu$, then we call $\mathbf{E}$ an *elementary matrix*. In particular we shall write $\mathbf{E}_{ij}^{mn}$ to indicate an $m \times n$ elementary matrix whose element in the ith row and the jth column is one. We may write, for fixed i and j,

$$\mathbf{E}_{ij}^{mn}=\|\delta_{\mu i}\delta_{\nu j}\|_{\mu\nu} \;, \quad 1 \leqq \mu \leqq m \;, \quad 1 \leqq \nu \leqq n \;. \tag{1}$$

The transpose of (1) is

$$(\mathbf{E}_{ij}^{mn})'=\mathbf{E}_{ji}^{nm} \;. \tag{2}$$

In particular

$$\mathbf{E}_{1j}^{1n}=\{\delta_{1j}, \cdots ,\delta_{nj}\} \tag{3}$$

is an n-dimensional row vector whose jth component is $+1$, and the remaining $n-1$ components are all zero. Similarly $\mathbf{E}_{i1}^{m1}$ is an m-dimensional column vector whose only nonzero entry is the ith component, and this nonzero entry is $+1$.

We also have the very useful identities

$$\mathbf{E}_{\mu\nu}^{mn}=\sum_{i=1}^{m}\sum_{j=1}^{n} \delta_{i\mu}\delta_{j\nu}\, \mathbf{E}_{ij}^{mn} \tag{4}$$

for any μ and ν; and if $m=n$,

$$\sum_{i=1}^{m} \mathbf{E}_{ii}^{mm}=\mathbf{I}_m \;. \tag{5}$$

For square elementary matrices we frequently omit the superscripts, that is, we write

$$\mathbf{E}_{ij}=\mathbf{E}_{ij}^{mm} \;, \tag{6}$$

it being understood from the context that $\partial\mathbf{E}_{ij}=m$.

By (1), the product of two compatible elementary matrices is

$$\mathbf{E}_{ij}^{mn}\mathbf{E}_{\mu\nu}^{nq}=\delta_{j\mu}\mathbf{E}_{i\nu}^{mq} \;, \tag{7}$$

which is an $m \times q$ elementary matrix if $j=\mu$. The elementary matrices may be used to "pick out" an element of an arbitrary matrix. For suppose $\mathbf{G}=\|g_{ij}\|$ is a $p \times q$ matrix. Then $g_{ij}\mathbf{E}_{ij}^{pq}$ has g_{ij} in its ith row and jth column and zeros elsewhere. Hence

$$\sum_{i=1}^{p}\sum_{j=1}^{q} g_{ij}\mathbf{E}_{ij}^{pq}=\mathbf{G} \;. \tag{8}$$

Some matrix manipulations also show that

$$\mathbf{E}_{ij}^{mp}\mathbf{G}\mathbf{E}_{\mu\nu}^{qn} = g_{j\mu}\mathbf{E}_{i\nu}^{mn} \quad . \tag{9}$$

That is, (9) represents an $m \times n$ matrix with $g_{j\mu}$ in its ith row and νth column.

Since $\mathbf{E}_{i1}^{m1}$ and $\mathbf{E}_{1j}^{1n}$ are vectors, we have

$$\mathbf{E}_{i1}^{m1}\mathbf{E}_{1j}^{1n} = \mathbf{E}_{ij}^{mn} \quad , \tag{10}$$

which is a special case of (7). If $m = n$, then

$$\mathbf{E}_{1j}^{1n}\ \mathbf{E}_{i1}^{n1} = \delta_{ij}$$

—which is a scalar.

Let

$$\mathcal{G} = \|\mathbf{G}_{\mu\nu}\| \quad , \quad 1 \leqq \mu \leqq m \quad , \quad 1 \leqq \nu \leqq n \quad ,$$

where the $\mathbf{G}_{\mu\nu}$ are $p \times q$ matrices. Then $\mathcal{G}$ is a partitioned $mp \times nq$ dimensional matrix of $m \times n$ blocks, each block being $p \times q$ dimensional. If $\mathbf{A}$ is an $r \times p$ matrix and $\mathbf{B}$ a $q \times s$ matrix, then

$$\|\mathbf{A}\mathbf{G}_{\mu\nu}\mathbf{B}\| = (\mathbf{A} \otimes \mathbf{I}_m)\mathcal{G}(\mathbf{B} \otimes \mathbf{I}_n) \tag{11}$$

is an $rm \times sn$ dimensional matrix. We also see that

$$\|\mathbf{C} \otimes \mathbf{G}_{\mu\nu}\| = \mathbf{C} \otimes \mathcal{G} \tag{12}$$

for any matrix $\mathbf{C}$.

Using (1), the Kronecker product of two elementary matrices is

$$\mathbf{E}_{ij}^{mn} \otimes \mathbf{E}_{\mu\nu}^{pq} = \|\delta_{\alpha\mu}\delta_{\beta\nu}\mathbf{E}_{ij}^{mn}\|_{\alpha\beta} \quad , \quad 1 \leqq \alpha \leqq p \quad , \quad 1 \leqq \beta \leqq q \quad , \tag{13}$$

which is an $mp \times nq$ dimensional matrix consisting of $p \times q$ blocks, each block being $m \times n$ dimensional. We see that (13) is an elementary matrix, the element in the ith row and jth column of the μ, ν block being $+1$. Thus we may write (13) in the alternative form

$$\mathbf{E}_{ij}^{mn} \otimes \mathbf{E}_{\mu\nu}^{pq} = \mathbf{E}_{(\mu-1)m+i,(\nu-1)n+j}^{mp,nq} \quad . \tag{14}$$

If we define $\mathbf{Q}$ as the $m^2 \times n^2$ partitioned matrix

$$\mathbf{Q} = \|\mathbf{E}_{\mu\nu}^{mn}\|_{\mu\nu} = \sum_{i=1}^{m} \sum_{j=1}^{n} \|\delta_{i\mu}\delta_{j\nu}\mathbf{E}_{ij}^{mn}\|_{\mu\nu} \tag{15a}$$

[see (4)] then by (13)

$$\mathbf{Q} = \sum_{i=1}^{m} \sum_{j=1}^{n} \mathbf{E}_{ij}^{mn} \otimes \mathbf{E}_{ij}^{mn} \ . \tag{15b}$$

From (2) the transpose of $\mathbf{Q}$ is

$$\mathbf{Q}' = \sum_{i=1}^{m} \sum_{j=1}^{n} \mathbf{E}_{ji}^{nm} \otimes \mathbf{E}_{ji}^{nm} \tag{16}$$

and by (7)

$$\mathbf{Q}'\mathbf{Q} = \sum_{i,\mu}^{m} \sum_{j,\nu}^{n} \delta_{i\mu}\mathbf{E}_{j\nu} \otimes \delta_{i\mu}\mathbf{E}_{j\nu}$$

$$= m \sum_{j,\nu}^{n} \mathbf{E}_{j\nu} \otimes \mathbf{E}_{j\nu} \ .$$

If $m = n$, then $\mathbf{Q} = \mathbf{Q}'$ and $\mathbf{Q}^2 = m\mathbf{Q}$.

A particularly useful matrix is the $mn \times mn$ partitioned matrix $\mathbf{P}$ defined as

$$\begin{aligned} \mathbf{P} = \|\mathbf{E}_{\mu\nu}^{mn}\|_{\nu\mu} &= \sum_{i=1}^{m} \sum_{j=1}^{n} \mathbf{E}_{ij}^{mn} \otimes \mathbf{E}_{ji}^{nm} \ . \\ &= \sum_{i=1}^{m} \sum_{j=1}^{n} \mathbf{E}_{ij}^{mn} \otimes (\mathbf{E}_{ij}^{mn})' \end{aligned} \tag{17}$$

(compare with the definition of $\mathbf{Q}$). We assert that $\mathbf{P}$ is an orthogonal matrix. First we may write

$$\mathbf{P}'\mathbf{P} = \sum_{i,\mu}^{m} \sum_{j,\nu}^{n} \mathbf{E}_{ji}^{nm}\mathbf{E}_{\mu\nu}^{mn} \otimes \mathbf{E}_{ij}^{mn}\mathbf{E}_{\nu\mu}^{nm}$$

$$= \sum_{j,\nu}^{n} \delta_{j\nu}\mathbf{E}_{j\nu} \otimes \sum_{i,\mu}^{m} \delta_{i\mu}\mathbf{E}_{i\mu}$$

by (7). And hence by (5)

$$\mathbf{P}'\mathbf{P} = \mathbf{I}_n \otimes \mathbf{I}_m = \mathbf{I}_{mn} \ . \tag{18}$$

If $m = n$, then $\mathbf{P}$ is symmetric.

Let $\mathbf{A} = \|a_{ij}\|$ be an $m \times m$ matrix and $\mathbf{B} = \|b_{ij}\|$ an $n \times n$ matrix. We shall show that

$$\mathbf{P}'(\mathbf{A} \otimes \mathbf{B})\mathbf{P} = \mathbf{B} \otimes \mathbf{A} \ . \tag{19}$$

That is, we have interchanged the order of the terms in the Kronecker product. To prove (19) we write

$$\mathbf{P}'(\mathbf{A} \otimes \mathbf{B})\mathbf{P} = \sum_{i,\mu}^{m} \sum_{j,\nu}^{n} (\mathbf{E}_{ji}^{nm} \otimes \mathbf{E}_{ij}^{mn})(\mathbf{A} \otimes \mathbf{B})(\mathbf{E}_{\mu\nu}^{mn} \otimes \mathbf{E}_{\nu\mu}^{nm})$$

$$= \sum_{i,\mu}^{m} \sum_{j,\nu}^{n} (\mathbf{E}_{ji}^{nm}\mathbf{A}\mathbf{E}_{\mu\nu}^{mn}) \otimes (\mathbf{E}_{ij}^{mn}\mathbf{B}\mathbf{E}_{\nu\mu}^{nm})$$

and by (9) and (13)

$$\mathbf{P}'(\mathbf{A} \otimes \mathbf{B})\mathbf{P} = \sum_{i,\mu}^{m} \sum_{j,\nu}^{n} a_{i\mu} b_{j\nu} (\mathbf{E}_{j\nu} \otimes \mathbf{E}_{i\mu})$$

$$= \sum_{i,\mu}^{m} \sum_{j,\nu}^{n} a_{i\mu} b_{j\nu} \|\delta_{\alpha i}\delta_{\beta\mu}\mathbf{E}_{j\nu}\|_{\alpha\beta}$$

$$= \sum_{i,\mu}^{m} a_{i\mu} \left\| \delta_{\alpha i}\delta_{\beta\mu} \sum_{j,\nu}^{n} b_{j\nu}\mathbf{E}_{j\nu} \right\|_{\alpha\beta} .$$

Thus by (8)

$$\mathbf{P}'(\mathbf{A} \otimes \mathbf{B})\mathbf{P} = \sum_{i,\mu}^{m} a_{i\mu} \|\delta_{\alpha i}\delta_{\beta\mu}\mathbf{B}\|_{\alpha\beta}$$

$$= \|a_{\alpha\beta}\mathbf{B}\|$$

$$= \mathbf{B} \otimes \mathbf{A}$$

—which is (19).

11. MATRIX INVERSION III—Iterative inverse of the sum of matrices

If $\mathbf{G}$ and $\mathbf{G}+\mathbf{H}$ are nonsingular, where $\mathbf{H}$ has rank one, then the inverse of $\mathbf{G}+\mathbf{H}$ is given by Corollary 1 of Section 4. If $\rho\mathbf{H}=2$, we still can find $(\mathbf{G}+\mathbf{H})^{-1}$ by using (4.4), and, in fact, if $\mathbf{H}$ has any rank, Theorem 2 of Section 4 is applicable. However, we now would like to consider an alternative, iterative, approach.

To introduce the technique, let us suppose that $\mathbf{H}$ has rank two and that we may write $\mathbf{H}$ in the form

$$\mathbf{H}=\mathbf{H}_1+\mathbf{H}_2 \tag{1}$$

where $\mathbf{H}_1$ and $\mathbf{H}_2$ both have rank one. Suppose further that

$$\mathbf{C}_2=\mathbf{G}+\mathbf{H}_1 \tag{2a}$$

and

$$\mathbf{C}_3=\mathbf{C}_2+\mathbf{H}_2=\mathbf{G}+\mathbf{H}_1+\mathbf{H}_2=\mathbf{G}+\mathbf{H} \tag{2b}$$

are nonsingular. Then by Corollary 4.1,

$$\mathbf{C}_2^{-1}=\mathbf{G}^{-1}-\frac{1}{1+\operatorname{tr}\mathbf{H}_1\mathbf{G}^{-1}}\mathbf{G}^{-1}\mathbf{H}_1\mathbf{G}^{-1} \tag{3a}$$

and

$$\mathbf{C}_3^{-1}=\mathbf{C}_2^{-1}-\frac{1}{1+\operatorname{tr}\mathbf{H}_2\mathbf{C}_2^{-1}}\mathbf{C}_2^{-1}\mathbf{H}_2\mathbf{C}_2^{-1}\ . \tag{3b}$$

Thus we have found, in an iterative fashion, the inverse of $\mathbf{G}+\mathbf{H}$ in terms of $\mathbf{G}^{-1}$; we merely substitute $\mathbf{C}_2^{-1}$ as given by (3a) into (3b).

Before generalizing this result to matrices $\mathbf{H}$ of arbitrary rank, let us consider a simple numerical example. Suppose

$$\mathbf{R}=\begin{bmatrix} 2 & 2 & 3 \\ 2 & 4 & 4 \\ 3 & 4 & 6 \end{bmatrix} \tag{4}$$

(a nonsingular matrix) and we write $\mathbf{R}=\mathbf{G}+\mathbf{H}$ where $\mathbf{G}$ is the 3×3 identity matrix and

$$\mathbf{H}=\begin{bmatrix} 1 & 2 & 3 \\ 2 & 3 & 4 \\ 3 & 4 & 5 \end{bmatrix}.$$

(The matrix $\mathbf{H}$ is a special case of an arithmetic matrix. Such matrices will be discussed more fully in Section 14.) Now $\rho\mathbf{H}=2$ and if we write $\mathbf{H}=\mathbf{H}_1+\mathbf{H}_2$ where

$$\mathbf{H}_1=\begin{bmatrix}0 & 1 & 2\\ 0 & 1 & 2\\ 0 & 1 & 2\end{bmatrix}$$

and

$$\mathbf{H}_2=\begin{bmatrix}1 & 1 & 1\\ 2 & 2 & 2\\ 3 & 3 & 3\end{bmatrix}$$

we see that $\mathbf{H}_1$ and $\mathbf{H}_2$ each have rank one. Since tr $\mathbf{H}_1\mathbf{G}^{-1}=\text{tr }\mathbf{H}_1=3$ we have [see (3a)]

$$\begin{aligned}\mathbf{C}_2^{-1}&=\mathbf{I}-\tfrac{1}{4}\,\mathbf{H}_1\\ &=\frac{1}{4}\begin{bmatrix}4 & -1 & -2\\ 0 & 3 & -2\\ 0 & -1 & 2\end{bmatrix}.\end{aligned}$$

Also,

$$\mathbf{H}_2\mathbf{C}_2^{-1}=\frac{1}{4}\begin{bmatrix}4 & 1 & -2\\ 8 & 2 & -4\\ 12 & 3 & -6\end{bmatrix},$$

tr $\mathbf{H}_2\mathbf{C}_2^{-1}=0$, and by (3b)

$$\begin{aligned}\mathbf{C}_3^{-1}=\mathbf{R}^{-1}&=\mathbf{C}_2^{-1}-\mathbf{C}_2^{-1}\mathbf{H}_2\mathbf{C}_2^{-1}\\ &=\mathbf{C}_2^{-1}(\mathbf{I}-\mathbf{H}_2\mathbf{C}_2^{-1})\\ &=\frac{1}{4}\begin{bmatrix}8 & 0 & -4\\ 0 & 3 & -2\\ -4 & -2 & 4\end{bmatrix}.\end{aligned}$$

Thus we have found, iteratively, the inverse of (4).

Now suppose $\mathbf{G}$ and $\mathbf{G}+\mathbf{H}$ are nonsingular and that $\mathbf{H}$ has positive rank r. We wish to determine an iterative scheme for finding the inverse of $\mathbf{G}+\mathbf{H}$. Our basic idea, as illustrated above for $r=2$, is to decompose $\mathbf{H}$ into the sum of matrices of rank one, say

$$\mathbf{H}=\mathbf{H}_1+\mathbf{H}_2+\cdots+\mathbf{H}_r \tag{5}$$

[see (1)] and iteratively apply Corollary 4.1. (See [Miller, 1981], [Wilf].) However, if we are to recursively apply Corollary 4.1 we need

$$\mathbf{C}_{k+1} = \mathbf{G} + \mathbf{H}_1 + \cdots + \mathbf{H}_k \tag{6}$$

[see (2)] to be nonsingular for $k = 1, \cdots, r$. (Note that $\mathbf{C}_1 = \mathbf{G}$ and $\mathbf{C}_{r+1} = \mathbf{G} + \mathbf{H}$.) Hence our first task is to show that a decomposition such as (5) always exists, and second we must show that the $\mathbf{H}_k$ may be chosen so that (6) holds. We do this in Lemmas 1 and 2 below. Then we shall be in a position to prove our main result, Theorem 1.

Lemma 1. Let $\mathbf{H}$ be an $n \times n$ matrix of rank r. Then $\mathbf{H}$ may be expressed in the form

$$\mathbf{H} = \mathbf{E}_1 + \mathbf{E}_2 + \cdots + \mathbf{E}_r \tag{7}$$

where each $\mathbf{E}_k$ has rank one.

Proof. Let $\partial \mathbf{E}_{ij} = n$ where $\mathbf{E}_{ij}$ is an elementary matrix (see Section 10). Let $\lambda_1, \cdots, \lambda_s$ $(s \leqq r)$ be the nonzero eigenvalues of $\mathbf{H}$. Then we may write the Jordan normal form $\mathbf{J}$ of $\mathbf{H}$ as

$$\mathbf{J} = (\lambda_1 \mathbf{E}_{11} + \epsilon \mathbf{E}_{12}) + \cdots + (\lambda_s \mathbf{E}_{ss} + \epsilon \mathbf{E}_{s,s+1}) + \mathbf{E}_{s+1,s+2} + \cdots + \mathbf{E}_{r,r+1}$$

where the generic ϵ is zero or one. ■

The decomposition (7) of a matrix of rank r into the sum of r matrices of rank one is not unique [Halmos, p. 93].

Lemma 2. Let $\mathbf{G}$ and $\mathbf{G} + \mathbf{H}$ be nonsingular and let $\mathbf{H}$ have positive rank r. Then there exists a decomposition

$$\mathbf{H} = \mathbf{H}_1 + \mathbf{H}_2 + \cdots + \mathbf{H}_r \tag{8}$$

of $\mathbf{H}$ such that $\rho \mathbf{H}_k = 1, \quad 1 \leqq k \leqq r$, and

$$\mathbf{C}_{k+1} = \mathbf{G} + \mathbf{H}_1 + \cdots + \mathbf{H}_k \;, \; k = 1, \cdots, r \tag{9}$$

is nonsingular.

Proof. Note that $\mathbf{C}_{r+1} = \mathbf{G} + \mathbf{H}$. We define $\mathbf{C}_1$ as $\mathbf{G}$. Now $\mathbf{G} + \mathbf{H} = (\mathbf{I} + \mathbf{H}\mathbf{G}^{-1})\mathbf{G}$ is nonsingular. Hence no eigenvalue of $\mathbf{H}\mathbf{G}^{-1}$ can be -1. Let $\mathbf{Q}$ be a nonsingular matrix such that $\mathbf{J} = \mathbf{Q}(\mathbf{H}\mathbf{G}^{-1})\mathbf{Q}^{-1}$ is the Jordan normal form of $\mathbf{H}\mathbf{G}^{-1}$, and write

$$\mathbf{J} = \mathbf{F}_1 + \mathbf{F}_2 + \cdots + \mathbf{F}_r$$

where the kth row of $\mathbf{F}_k$ is the kth row of $\mathbf{J}$ and the remaining rows of $\mathbf{F}_k$ all are zero. Then every $\mathbf{F}_k$ has rank one and $\mathbf{I}+\mathbf{F}_1+\cdots+\mathbf{F}_k$ is nonsingular for $k=1,\cdots,r$. Thus

$$[\mathbf{Q}^{-1}(\mathbf{I}+\mathbf{F}_1+\cdots+\mathbf{F}_k)\mathbf{Q}]\mathbf{G}=\mathbf{G}+\mathbf{Q}^{-1}\mathbf{F}_1\mathbf{Q}\mathbf{G}+\cdots+\mathbf{Q}^{-1}\mathbf{F}_k\mathbf{Q}\mathbf{G}$$

is nonsingular for $k=1,\cdots,r$, and $\mathbf{H}_j=\mathbf{Q}^{-1}\mathbf{F}_j\mathbf{Q}\mathbf{G}$ has rank one. ∎

We are now in a position to prove our main result,

Theorem 1. Let $\mathbf{G}$ and $\mathbf{G}+\mathbf{H}$ be nonsingular matrices and let $\rho\mathbf{H}=r>0$. Let $\mathbf{H}=\mathbf{H}_1+\cdots+\mathbf{H}_r$ where $\rho\mathbf{H}_k=1$, $1\leqq k\leqq r$, and $\mathbf{C}_{k+1}=\mathbf{G}+\mathbf{H}_1+\cdots+\mathbf{H}_k$ is nonsingular for $k=1,\cdots,r$. Then with $\mathbf{C}_1=\mathbf{G}$,

$$\mathbf{C}_{k+1}^{-1}=\mathbf{C}_k^{-1}-\nu_k\mathbf{C}_k^{-1}\mathbf{H}_k\mathbf{C}_k^{-1}\ ,\quad k=1,\cdots,r \tag{10}$$

where

$$\nu_k^{-1}=1+\operatorname{tr}\mathbf{C}_k^{-1}\mathbf{H}_k\ ,\quad 1\leqq k\leqq r\ . \tag{11}$$

In particular

$$(\mathbf{G}+\mathbf{H})^{-1}=\mathbf{C}_r^{-1}-\nu_r\mathbf{C}_r^{-1}\mathbf{H}_r\mathbf{C}_r^{-1}\ . \tag{12}$$

Proof. Since $\mathbf{C}_2=\mathbf{C}_1+\mathbf{H}_1=\mathbf{G}+\mathbf{H}_1$ and $\mathbf{G}$ and $\mathbf{C}_2$ are nonsingular, Corollary 4.1 implies that $\nu_1\neq 0$ and

$$\mathbf{C}_2^{-1}=\mathbf{G}^{-1}-\nu_1\mathbf{G}^{-1}\mathbf{H}_1\mathbf{G}^{-1}\ .$$

Inductively we then establish (10) and hence (12). ∎

As a corollary we have

Corollary 1. Under the hypotheses of Theorem 1,

$$\det(\mathbf{G}+\mathbf{H})=(\nu_1\nu_2\cdots\nu_r)^{-1}\det\mathbf{G}\ . \tag{13}$$

Proof. From (9)

$$\mathbf{C}_{k+1}=\mathbf{C}_k+\mathbf{H}_k=\mathbf{C}_k(\mathbf{I}+\mathbf{C}_k^{-1}\mathbf{H}_k)$$

and

$$\det\mathbf{C}_{k+1}=\nu_k^{-1}\det\mathbf{C}_k\ .$$

Inductively we establish (13). ∎

In many practical problems **G** is positive definite and **H** is diagonal and nonnegative definite. In such cases $\mathbf{G}+\mathbf{H}$ is always nonsingular and the decomposition (8) of **H** is trivial. The recursive form is especially convenient for computer utilization.

12. MATRIX INVERSION IV—Inversion of infinite stationary matrices

Despite all our previous efforts, the problem of inverting an arbitrary nonsingular matrix is intractable. If the dimension of the matrix is small, a direct attack is feasible; and if the elements of the matrix are simply numbers, a computer approach is possible. However, even in this latter case we sometimes encounter difficulties. For example, the determinant of the matrix $\mathbf{M}=\|m^{2n}\|_{mn}$ exceeds 10^{268} for $\partial\mathbf{M} \geqq 16$ (see Section 14); and we are in danger of having the computer flash "algorithmically singular". In this section we shall consider a class of matrices (stationary matrices) which have the property that their inverses may be found approximately when the dimension of the matrix is large.

Let $\{r(n)|n=0,1,2,\cdots\}$ be a sequence of numbers and let $\mathbf{R}=\|r(|m-n|)\|_{mn}$ be a square matrix of arbitrary dimension. If $\mathbf{R}$ is nonnegative definite we call $\mathbf{R}$ a *stationary matrix*. The inverse of a nonsingular stationary matrix is not necessarily stationary. The techniques we shall employ to find the inverse of $\mathbf{R}$ when $\mathbf{R}$ is positive definite and $\partial\mathbf{R}$ is large require additional assumptions on the sequence $\{r(n)\}$. Consider then the Fourier series

$$\sum_{m=-\infty}^{\infty} r(m)e^{-i2\pi\mu m} \tag{1}$$

where we define $r(-m)$ as $r(m)$ for m positive. If

$$\sum_{m=1}^{\infty} |r(m)| < \infty \ , \tag{2}$$

then by the Weierstrass M-test (see Appendix B), the infinite series of (1) converges uniformly and absolutely in any interval of the μ-axis. Thus (1) defines a continuous function, say $W(\mu)$, and W is even and nonnegative. Also, we may multiply (1) by $e^{i2\pi\mu n}$ and integrate term by term from $-\frac{1}{2}$ to $+\frac{1}{2}$ to obtain

$$r(n)=r(-n)=\int_{-\frac{1}{2}}^{\frac{1}{2}} W(\mu)e^{i2\pi n\mu}d\mu \ , \quad n=0,1,\cdots \ . \tag{3}$$

Now suppose W is bounded away from zero on $[-\frac{1}{2},\frac{1}{2}]$. Then

$$s(n)=s(-n)=\int_{-\frac{1}{2}}^{\frac{1}{2}} \frac{1}{W(\nu)}\, e^{i2\pi n\nu}d\nu \ , \quad n=0,1,\cdots \tag{4}$$

also exists, and we may consider the square matrix $\mathbf{S}=\|s(m-n)\|$ of arbitrary dimension. We shall show that if $\partial\mathbf{R}=\partial\mathbf{S}=M$, then as M increases without limit, **RS** approaches the identity matrix.

A precise formulation of this result will be given in Theorem 1 below. Before doing this, however, let us indicate the motivation that led to the above definition of W. The origin of the term stationary matrix has its roots in the theory of time series (see Appendix A). Let $\{y_t|t=\cdots,-1,0,+1,\cdots\}$ be a sequence of random variables. Let $\mathcal{E}y_t=0$ and

$$\mathcal{E}y_t y_\tau = r(t-\tau) \ .$$

Then $\{y_t\}$ is a wide-sense stationary stochastic process. The covariance function $r(t-\tau)$ is nonnegative definite and an even function, $r(t-\tau)=r(\tau-t)$. A stochastic process with a discrete time parameter which is stationary in the wide-sense defines a sequence of covariances $r(0)$, $r(1)$, $\cdots$. Let

$$W(\mu) \sim \sum_{m=-\infty}^{\infty} r(m)e^{-i2\pi\mu m} \tag{5}$$

[see (1)]. If the series on the right-hand side of (5) converges, then we call W the *spectral density function* of the time series. The spectral density function is nonnegative and even.

We now state and prove our basic theorem.

Theorem 1. Let $\{r(n)\}$ be a covariance sequence with corresponding spectral density W. Let W be continuously differentiable and bounded away from zero on $[-\frac{1}{2},\frac{1}{2}]$. Let $\mathbf{R}_N=\|r(m-n)\|$ and $\mathbf{S}_N=\|s(m-n)\|$ be $(2N+1)\times(2N+1)$ dimensional matrices where r and s are given by (3) and (4) respectively. Let

$$\mathbf{R}_N\mathbf{S}_N=\|\sigma_{mn}^{(N)}\| \ .$$

Then

$$\lim_{N\to\infty}\sigma_{mn}^{(N)}=\delta_{mn} \ .$$

Proof. Let us write

$$\mathbf{R}_N=\int_{-\frac{1}{2}}^{\frac{1}{2}} W(\mu)\mathbf{E}_N(\mu)d\mu$$

and

$$\mathbf{S}_N = \int_{-\frac{1}{2}}^{\frac{1}{2}} \frac{1}{W(\nu)} \mathbf{E}_N(\nu) d\nu$$

where $\mathbf{E}_N(\lambda) = \|e^{-i2\pi(m-n)\lambda}\|_{mn}$ and $\partial\mathbf{E}_N = 2N+1$. Then

$$\mathbf{R}_N\mathbf{S}_N = \int_{-\frac{1}{2}}^{\frac{1}{2}} \int_{-\frac{1}{2}}^{\frac{1}{2}} \frac{W(\mu)}{W(\nu)} \mathbf{E}_N(\mu)\mathbf{E}_N(\nu) d\mu d\nu \ .$$

If we let

$$\mathbf{E}_N(\mu)\mathbf{E}_N(\nu) = \|f_{mn}^{(N)}\|_{mn} \ ,$$

then

$$f_{mn}^{(N)} = \sum_{k=-N}^{N} e^{-i2\pi(m-k)\mu} e^{-i2\pi(k-n)\nu}$$

$$= e^{-i2\pi(m\mu - n\nu)} \frac{\sin \pi(\mu-\nu)(2N+1)}{\sin \pi(\mu-\nu)} \ .$$

Thus we may write (with $M = 2N+1$)

$$\mathbf{R}_N\mathbf{S}_N = \int_{-\frac{1}{2}}^{\frac{1}{2}} \int_{-\frac{1}{2}}^{\frac{1}{2}} \frac{W(\mu)}{W(\nu)} \frac{\sin \pi(\mu-\nu)M}{\sin \pi(\mu-\nu)} \|e^{-i2\pi(m\mu-n\nu)}\| d\mu d\nu \ . \tag{6}$$

The integrand in (6) is jointly continuous in μ and ν by hypothesis. Make the change of variable $x = \mu - \nu$, $y = \nu$. Then (6) becomes

$$\sigma_{mn}^{(N)} = \int_0^1 g(x) \frac{\sin \pi M x}{\sin \pi x} dx \tag{7}$$

where

$$g(x) = 2\int_{-\frac{1}{2}}^{\frac{1}{2}-x} \frac{W(y+x)}{W(y)} \cos 2\pi[mx + (m-n)y] dy \ . \tag{8}$$

Now define

$$h(x) = \frac{x}{\sin \pi x} g(x) \ , \quad 0 < x < 1 \ .$$

The joint continuity of the integrand of g [see (8)] implies that

$$\lim_{x \to 0} h(x) = \frac{2}{\pi} \delta_{mn}$$

and

$$\lim_{x\to 1} h(x)=\frac{2}{\pi}(-1)^{m-n} .$$

Hence $h(x)$ is integrable on [0,1] and we may write (7) as

$$\sigma_{mn}^{(N)}=\int_0^1 h(x)\,\frac{\sin \pi Mx}{x}\,dx .$$

The continuity of the derivative of the integrand of g with respect to x implies that $g(x)$ is continuously differentiable. Hence $h'(0+)$ exists, and by the corollary to the Riemann-Lebesgue lemma (see Appendix B)

$$\lim_{N\to\infty} \sigma_{mn}^{(N)}=\frac{\pi}{2}\,h(0+)=\delta_{mn} . \quad \blacksquare$$

As a first application of the above theorem we shall consider the Markoffian matrix

$$\mathbf{R}_N=\|\lambda^{|m-n|}\| \ , \ 0<\lambda<1 . \tag{9}$$

Then [see (5)],

$$W(\mu)=\sum_{m=-\infty}^{\infty} \lambda^{|m|}e^{-i2\pi\mu m}$$

$$=\frac{1-\lambda^2}{1-2\lambda\cos 2\pi\mu+\lambda^2}$$

and from (4),

$$s(n)=\frac{1}{1-\lambda^2}\int_{-\frac{1}{2}}^{\frac{1}{2}}(1+\lambda^2-2\lambda\cos 2\pi\nu)e^{i2\pi n\nu}d\nu .$$

Hence

$$\begin{aligned} s(0)&=\frac{1+\lambda^2}{1-\lambda^2}\\ s(\pm 1)&=\frac{-\lambda}{1-\lambda^2} \\ s(n)&=0 \qquad , |n|>1 . \end{aligned} \tag{10}$$

Later (see Section 14) we shall find the exact inverse of $\mathbf{R}_N$ for N finite.

Let us consider a variation of this problem, namely, the sum of a

Markoffian matrix and a scalar multiple of the identity matrix:

$$\mathbf{R}_N = \|\lambda^{|m-n|}\| + \gamma \mathbf{I}_{2N+1}\ , \quad 0 < \lambda < 1\ , \quad \gamma > 0\ . \tag{11}$$

This matrix occurs in various practical problems. If $\gamma = 0$, it reduces to (9). Again by (5)

$$W(\mu) = \gamma + \frac{1-\lambda^2}{1-2\lambda\cos 2\pi\mu + \lambda^2}$$

and

$$s(n) = \int_{-\frac{1}{2}}^{\frac{1}{2}} \frac{1-2\lambda\cos 2\pi\nu + \lambda^2}{\gamma[1-2\lambda\cos 2\pi\nu+\lambda^2] + (1-\lambda^2)}\, e^{i2\pi n\nu} d\nu\ . \tag{12}$$

Since $s(n) = s(-n)$ we need concern ourselves only with the case $n \geqq 0$. To evaluate (12) we find it convenient to transform (12) into a contour integral by the change of variable

$$z = e^{i2\pi\nu}\ .$$

Then

$$s(n) = \frac{1}{2\pi i}\int_\Gamma \frac{[\lambda z^2 - (1+\lambda^2)z + \lambda]z^{n-1}}{\lambda\gamma z^2 - [\gamma(1+\lambda^2) + (1-\lambda^2)]z + \lambda\gamma}\, dz\ , \quad n \geqq 0 \tag{13}$$

where Γ is the contour illustrated in Fig. 2 below, that is, the unit circle with center at the origin, traversed in a counterclockwise fashion.

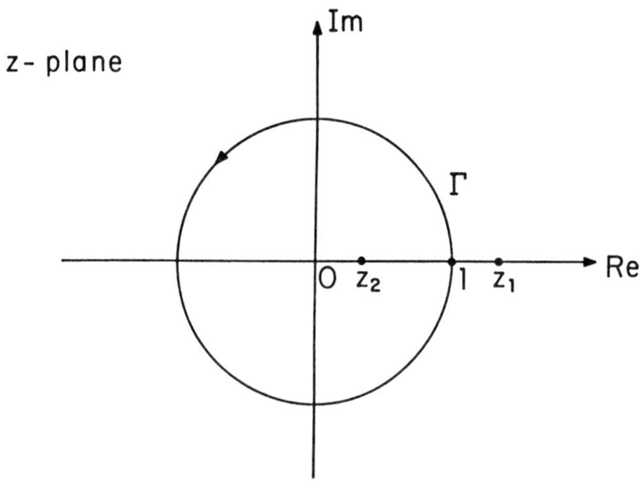

Figure 2

Let

$$g(z)=\lambda\gamma z^2-[\gamma(1+\lambda^2)+(1-\lambda^2)]z+\lambda\gamma$$

be the denominator of the integrand of (13). Then poles of the integrand occur at the roots of the equation $g(z)=0$. Both these roots are real. Let z_1 be the larger and z_2 the smaller. Then it is easy to see that

$$z_1 > 1 \qquad \text{and} \qquad 0 < z_2 < 1 \ .$$

Thus we may write (13) as

$$s(n)=\frac{1}{2\pi i\gamma}\int_\Gamma \frac{[\phi(z)-\psi(z)]z^n}{z-z_2}\,dz \ , \ n \geqq 0$$

where

$$\phi(z)=\frac{z+z^{-1}}{z-z_1}$$

and

$$\psi(z)=\frac{\lambda+\lambda^{-1}}{z-z_1} \ .$$

From the residue theorem (see Appendix B)

$$s(n)=\frac{g(\lambda)z_2^{|n|}}{\lambda^2\gamma^2(z_1-z_2)} \qquad , \ n \neq 0 \tag{14a}$$

$$s(n)=\frac{1}{\gamma}+\frac{g(\lambda)}{\lambda^2\gamma^2(z_1-z_2)} \ , \ n=0 \ . \tag{14b}$$

To verify directly that $\mathbf{S}_N=\|s(m-n)\|$ is the inverse of $\mathbf{R}_N$, take the inner product of the pth row of $\mathbf{R}_N$ and the $(p+q)$th column of $\mathbf{S}_N$. There results

$$\sum_{k=-N}^{N} (\lambda^{|p-k|}+\gamma\delta_{pk})(Az_2^{|p+q-k|}+\gamma^{-1}\delta_{p+q,k}) \tag{15}$$

where

$$A=\frac{g(\lambda)}{\lambda^2\gamma^2(z_1-z_2)}=-\frac{1-\lambda^2}{\lambda\gamma^2(z_1-z_2)} \ .$$

Some arithmetic shows, that as N increases without limit, (15) is one if $q=0$ and is zero if $q \neq 0$.

As our third example, let $r(0)=1$, $r(\pm 1)=\lambda$, $r(n)=0$ for $|n| > 1$.

Then $\mathbf{R}_N = \|r(m-n)\|$ is a tridiagonal matrix and

$$W(\mu) = 1 + 2\lambda \cos 2\pi\mu \ .$$

A sufficient condition for $\mathbf{R}_N$ to be positive definite for all N is that $|\lambda| < \frac{1}{2}$. Under this condition $W(\mu)$ is bounded away from zero.

Let $\mathbf{S}_N = ||s(m-n)||$ where

$$s(n) = \int_{-\frac{1}{2}}^{\frac{1}{2}} \frac{1}{1+2\lambda \cos 2\pi\nu} e^{i2\pi n\nu} d\nu \ .$$

Then under the transformation $z = e^{i2\pi\nu}$ we may write $s(n)$ as the contour integral

$$s(n) = \frac{1}{2\pi i} \int_\Gamma \frac{z^n dz}{\lambda z^2 + z + \lambda}$$

where Γ is the same contour illustrated in Fig. 2.

If $\lambda = 0$, both $\mathbf{R}_N$ and $\mathbf{S}_N$ are the identity matrix $\mathbf{I}_{2N+1}$. Let us therefore assume that $\lambda \neq 0$. Let

$$\begin{aligned} g(z) &= \lambda z^2 + z + \lambda \\ &= \lambda(z - z_1)(z - z_2) \ . \end{aligned}$$

Then z_1 and z_2 are real and $|z_1| > 1$, $0 < |z_2| < 1$. Hence

$$s(n) = \frac{z_2^{|n|}}{\lambda(z_2 - z_1)} \ . \tag{16}$$

Later (see Section 14) we shall find the exact inverse of $\mathbf{R}_N$ for N finite.

13. THE DERIVATIVE OF A MATRIX

In Section 5 we defined the derivative of a scalar with respect to a vector and indicated some useful applications. We also remarked that we could have defined the derivative of a scalar with respect to a matrix. Now we shall define the derivative of a matrix with respect to a matrix—which subsumes all previously mentioned cases. This more general approach could not have been well exploited earlier since certain machinery (namely Kronecker products and elementary matrices) was not yet available.

Let

$$\mathbf{F} = \|f_{ij}\|$$

be a $p \times q$ matrix and

$$\mathbf{X} = \|x_{\mu\nu}\|$$

an $m \times n$ matrix. Let $x_{\mu\nu}$ be considered as mn independent variables, and let the f_{ij} be differentiable functions of the $x_{\mu\nu}$ in some region $\mathcal{D}$ of mn-dimensional Euclidean space. Then $\partial f_{ij}/\partial x_{\mu\nu}$ exists for all i, j, μ, ν and we say that $\mathbf{F}$ is a *differentiable matrix* of the matrix $\mathbf{X}$ on $\mathcal{D}$ (see [Brewer]). We define $\partial\mathbf{F}/\partial\mathbf{X}$ for $\mathbf{X} \in \mathcal{D}$ as

$$\frac{\partial \mathbf{F}}{\partial \mathbf{X}} = \left\| \frac{\partial \mathbf{F}}{\partial x_{\mu\nu}} \right\| . \tag{1}$$

Thus $\partial\mathbf{F}/\partial\mathbf{X}$ is a $pm \times qn$ dimensional partitioned matrix consisting of $m \times n$ blocks, each block being $p \times q$ dimensional.

Before embarking on the general theory, let us consider some simple examples. Suppose $m = n$ so that $\mathbf{X}$ is a square matrix. Then

$$\frac{\partial}{\partial \mathbf{X}} \det \mathbf{X} = \mathcal{X}$$

where $\mathcal{X} = \|X_{\mu\nu}\|$ is the matrix of cofactors $X_{\mu\nu}$ of $x_{\mu\nu}$.

If $n = 1$, so that $\mathbf{X}$ is an m-dimensional column vector, then direct calculations show that

$$\frac{\partial \mathbf{X}'}{\partial \mathbf{X}} = \frac{\partial \mathbf{X}}{\partial \mathbf{X}'} = \mathbf{I}_m , \tag{2}$$

and if $\mathbf{C}$ is an $r \times m$ matrix independent of $\mathbf{X}$,

$$\frac{\partial}{\partial \mathbf{X}'} \mathbf{CX} = \mathbf{C} \ . \tag{3}$$

Returning to the general case (where $\mathbf{X}$ is $m \times n$ dimensional), it is easy to show that

$$\left(\frac{\partial \mathbf{F}}{\partial \mathbf{X}}\right)' = \frac{\partial \mathbf{F}'}{\partial \mathbf{X}'} \ . \tag{4}$$

In particular

$$\frac{\partial \mathbf{X}}{\partial \mathbf{X}} = \|\mathbf{E}^{mn}_{\mu\nu}\|_{\mu\nu} = \mathbf{Q} \tag{5}$$

[see (10.15)] and

$$\frac{\partial \mathbf{X}'}{\partial \mathbf{X}} = \|\mathbf{E}^{nm}_{\nu\mu}\|_{\mu\nu} = \mathbf{P}' \tag{6}$$

[see (10.17)]. Equation (2) can, of course, formally be deduced from (6). For in this case $n=1$ and (6) becomes the $m \times m$ dimensional matrix $\|\mathbf{E}^{1m}_{1\mu}\|_{\mu 1}$,

$$\frac{\partial \mathbf{X}'}{\partial \mathbf{X}} = \|\mathbf{E}^{1m}_{1\mu}\|_{\mu 1} \ .$$

By (10.17), (9.8) and (10.10),

$$\begin{aligned}\|\mathbf{E}^{1m}_{1\mu}\|_{\mu 1} &= \sum_{i=1}^{m} \mathbf{E}^{1m}_{1i} \otimes \mathbf{E}^{m1}_{i1} \\ &= \sum_{i=1}^{m} \mathbf{E}^{m1}_{i1} \mathbf{E}^{1m}_{1i} \\ &= \sum_{i=1}^{m} \mathbf{E}^{mm}_{ii} \ .\end{aligned}$$

Thus

$$\frac{\partial \mathbf{X}'}{\partial \mathbf{X}} = \mathbf{I}_m$$

by (10.5)—which is (2).

After these preliminary remarks, let us attempt to develop a systematic calculus of differentiation with respect to a matrix. If $\mathbf{C}$ is a constant matrix,

$$\frac{\partial \mathbf{C}}{\partial \mathbf{X}}=\mathbf{0}\ ; \tag{7}$$

and if $\mathbf{F}$ and $\mathbf{G}$ are compatible differentiable matrices of $\mathbf{X}$ on $\mathcal{D}$, and if a and b are constant scalars, then

$$\frac{\partial}{\partial \mathbf{X}}(a\mathbf{F}+b\mathbf{G})=a\frac{\partial \mathbf{F}}{\partial \mathbf{X}}+b\frac{\partial \mathbf{G}}{\partial \mathbf{X}}\ . \tag{8}$$

Equations (7) and (8) are obvious consequences of the definition (1) of differentiation of a matrix with respect to a matrix.

We shall consider the product differentiation formulas for three cases, namely, the product of a scalar and a matrix, the product of two matrices, and the Kronecker product of two matrices. The formula for the derivative of the inverse of a matrix also will be derived.

Suppose f is a differentiable scalar function of $\mathbf{X}$, and $\mathbf{F}$ a differentiable matrix of $\mathbf{X}$. Then

$$\frac{\partial}{\partial x_{\mu\nu}}f\mathbf{F}=\frac{\partial f}{\partial x_{\mu\nu}}\mathbf{F}+f\frac{\partial \mathbf{F}}{\partial x_{\mu\nu}}$$

and hence

$$\frac{\partial}{\partial \mathbf{X}}f\mathbf{F}=\mathbf{F}\otimes\frac{\partial f}{\partial \mathbf{X}}+f\frac{\partial \mathbf{F}}{\partial \mathbf{X}}\ . \tag{9}$$

Also, if $\mathbf{F}$ and $\mathbf{G}$ are compatible differentiable matrices of $\mathbf{X}$, then

$$\frac{\partial}{\partial x_{\mu\nu}}\mathbf{F}\mathbf{G}=\frac{\partial \mathbf{F}}{\partial x_{\mu\nu}}\mathbf{G}+\mathbf{F}\frac{\partial \mathbf{G}}{\partial x_{\mu\nu}}$$

and by (10.11),

$$\frac{\partial}{\partial \mathbf{X}}\mathbf{F}\mathbf{G}=\frac{\partial \mathbf{F}}{\partial \mathbf{X}}(\mathbf{G}\otimes\mathbf{I}_n)+(\mathbf{F}\otimes\mathbf{I}_m)\frac{\partial \mathbf{G}}{\partial \mathbf{X}}\ . \tag{10}$$

As a corollary to (10), we see that if $\mathbf{F}$ is a constant matrix, say $\mathbf{C}$, then

$$\frac{\partial}{\partial \mathbf{X}}\mathbf{C}\mathbf{G}=(\mathbf{C}\otimes\mathbf{I}_m)\frac{\partial \mathbf{G}}{\partial \mathbf{X}}\ . \tag{11}$$

If $\mathbf{G}$ is a constant matrix, say $\mathbf{K}$, and $\mathbf{F}=\mathbf{X}'$, then (10) implies

$$\frac{\partial}{\partial \mathbf{X}}\mathbf{X}'\mathbf{K}=\frac{\partial \mathbf{X}'}{\partial \mathbf{X}}(\mathbf{K}\otimes\mathbf{I}_n) \tag{12}$$

and by (6),

$$\frac{\partial}{\partial \mathbf{X}} \mathbf{X}'\mathbf{K} = \mathbf{P}'(\mathbf{K} \otimes \mathbf{I}_n) \ .$$

Now let $n=1$ so that $\mathbf{X}$ is an m-dimensional vector. Then $\partial\mathbf{X}'/\partial\mathbf{X} = \mathbf{I}_m$ by (2) and (12) reduces to

$$\frac{\partial}{\partial \mathbf{X}} \mathbf{X}'\mathbf{K} = \mathbf{I}_m(\mathbf{K} \otimes \mathbf{I}_1) = \mathbf{I}_m\mathbf{K} = \mathbf{K} \tag{13}$$

by (9.10). Letting $\mathbf{K} = \mathbf{C}'$ we see that (13) is the transpose of (3) [see (4)].

Before considering the derivative of the Kronecker product, let us find the derivative of the inverse of a matrix. Suppose, then, that $\mathbf{F}$ is a nonsingular differentiable matrix of $\mathbf{X}$ on $\mathfrak{D}$. Then the identity

$$\frac{\partial}{\partial \mathbf{X}} \mathbf{F}\mathbf{F}^{-1} = \mathbf{0}$$

together with (10) imply

$$\frac{\partial \mathbf{F}^{-1}}{\partial \mathbf{X}} = -(\mathbf{F}^{-1} \otimes \mathbf{I}_m) \frac{\partial \mathbf{F}}{\partial \mathbf{X}} (\mathbf{F}^{-1} \otimes \mathbf{I}_n) \ . \tag{14}$$

Now let $\mathbf{F}$ and $\mathbf{G}$ be square matrices (not necessarily of the same dimension). We shall assume that both $\mathbf{F}$ and $\mathbf{G}$ are differentiable matrices of the matrix $\mathbf{X}$ on $\mathfrak{D}$. Then by (10.19)

$$\begin{aligned} \frac{\partial}{\partial x_{\mu\nu}} (\mathbf{F} \otimes \mathbf{G}) &= \mathbf{F} \otimes \frac{\partial \mathbf{G}}{\partial x_{\mu\nu}} + \frac{\partial \mathbf{F}}{\partial x_{\mu\nu}} \otimes \mathbf{G} \\ &= \mathbf{F} \otimes \frac{\partial \mathbf{G}}{\partial x_{\mu\nu}} + \mathbf{P}'\left(\mathbf{G} \otimes \frac{\partial \mathbf{F}}{\partial x_{\mu\nu}}\right)\mathbf{P} \ . \end{aligned}$$

With the aid of (10.11) we therefore may write

$$\frac{\partial}{\partial \mathbf{X}} (\mathbf{F} \otimes \mathbf{G}) = \mathbf{F} \otimes \frac{\partial \mathbf{G}}{\partial \mathbf{X}} + (\mathbf{P}' \otimes \mathbf{I}_m)\left(\mathbf{G} \otimes \frac{\partial \mathbf{F}}{\partial \mathbf{X}}\right)(\mathbf{P} \otimes \mathbf{I}_n) \ . \tag{15}$$

In the above two formulas, of course, $\partial\mathbf{P} = (\partial\mathbf{F})(\partial\mathbf{G})$.

14. MATRIX INVERSION V—Inverses of some special matrices

There exists a plethora of special matrices that occur in theoretical as well as applied problems. We may mention matrices associated with the names Toeplitz (= stationary), Jacobi (=tridiagonal), Markov, Gauss, Vandermonde, Hilbert. In this section we shall analyze these and other matrices.

(a) Markoffian matrices

The inverse $\mathbf{S}_N=\|s_{mn}\|$ of the $N\times N$ dimensional matrix

$$\mathbf{R}_N=\|\lambda^{|m-n|}\| \ , \ 0<\lambda<1 \tag{1}$$

is tridiagonal, as is its infinite extension [see (12.10)]. Explicitly,

$$s_{11}=s_{NN}=\frac{1}{1-\lambda^2}$$

$$s_{mm}=\frac{1+\lambda^2}{1-\lambda^2} \ , \ 1\neq m\neq N$$

$$s_{mn}=\frac{-\lambda}{1-\lambda^2} \ , \ |m-n|=1$$

and $s_{mn}=0$ otherwise. Note that except for s_{11} and s_{NN}, the infinite and finite Markoffian inverses have the same elements.

In Section (b) we shall find the inverse of the $N\times N$ dimensional tridiagonal stationary matrix

$$\mathbf{A}_N=\|\delta_{mn}+\lambda(\delta_{m,n+1}+\delta_{m,n-1})\| \tag{2a}$$

when $\mathbf{A}_N$ is nonsingular. Its infinite extension is of Markoffian character [see (12.16)].

Let λ_i with

$$\lambda_1>\lambda_2>\cdots>\lambda_N>0$$

be the eigenvalues of the positive definite Markoffian matrix (1). If $N=2$,

$$\lambda_1=1+\lambda$$

$$\lambda_2=1-\lambda \quad ;$$

if $N=3$,

$$\lambda_1=1+\tfrac{1}{2}\lambda(\lambda+\sqrt{8+\lambda^2})$$

$$\lambda_2=1-\lambda^2$$

$$\lambda_3=1+\tfrac{1}{2}\lambda(\lambda-\sqrt{8+\lambda^2}) \ ;$$

and if $N=4$,

$$\lambda_1 = 1 + \tfrac{1}{2}\lambda[(1+\lambda^2) + (1+\lambda)\sqrt{5-2\lambda+\lambda^2}]$$

$$\lambda_2 = 1 - \tfrac{1}{2}\lambda[(1+\lambda^2) - (1-\lambda)\sqrt{5+2\lambda+\lambda^2}]$$

$$\lambda_3 = 1 + \tfrac{1}{2}\lambda[(1+\lambda^2) - (1+\lambda)\sqrt{5-2\lambda+\lambda^2}]$$

$$\lambda_4 = 1 - \tfrac{1}{2}\lambda[(1+\lambda^2) + (1-\lambda)\sqrt{5+2\lambda+\lambda^2}] \; .$$

If N is very large, then

$$\lambda_1 = \frac{1+\lambda}{1-\lambda}$$

and

$$\lambda_N = \frac{1-\lambda}{1+\lambda} \; ,$$

see [Grenander and Szegö]. For various values of λ and N, the largest eigenvalue of $\mathbf{R}_N = \|\lambda^{|m-n|}\|$ is given in Table I below.

Largest Eigenvalue of $\mathbf{R}_N = \|\lambda^{|m-n|}\|$

N \ λ	0.5	0.6	0.7	0.8	0.9
2	1.5000	1.6000	1.7000	1.8000	1.9000
3	1.8431	2.0474	2.2648	2.4958	2.7407
4	2.0856	2.3870	2.7248	3.1032	3.5266
8	2.5716	3.1515	3.9001	4.8846	6.2030
16	2.8440	3.6582	4.8530	6.7269	9.9268
32	2.9524	3.8888	5.3745	8.0407	13.882
∞	3.0000	4.0000	5.6667	9.0000	19.000

Table I

(b) Tridiagonal stationary matrices

Let

$$\mathbf{A}_N = \|\lambda_{mn}\| \tag{2b}$$

where

$$\lambda_{mm} = 1$$

$$\lambda_{mn} = \lambda \ , \ |m-n| = 1 \ ,$$

and $\lambda_{mn} = 0$ otherwise. That is, (2b) is the tridiagonal matrix of (2a) written in less compact, but perhaps more convenient form. We shall assume that $\lambda \neq 0$. For if $\lambda = 0$, then $\mathbf{A}_N$ would be the $N \times N$ dimensional identity matrix.

By the Laplace development of a determinant

$$D_N = D_k D_{N-k} - \lambda^2 D_{k-1} D_{N-k-1} \ , \ 1 \leqq k \leqq N-1 \tag{3}$$

where

$$D_N = \det \mathbf{A}_N \ , \ N = 1,2, \cdots$$

and $D_0 = 1 = D_1$. In particular $D_2 = 1 - \lambda^2$ and

$$D_3 = 1 - 2\lambda^2 \ .$$

We see from D_2 and D_3 that D_N is not necessarily nonsingular unless restrictions are placed on λ. From Section 12 we know that $\mathbf{A}_N$ will be positive definite provided $|\lambda| \leqq \frac{1}{2}$. Thus $\mathbf{A}_N$ is always nonsingular for any N, if $|\lambda| \leqq \frac{1}{2}$.

A direct calculation now shows that if $\mathbf{A}_N$ is nonsingular,

$$b_{mn} = b_{nm} = (-\lambda)^{n-m} \frac{D_{N-n} D_{m-1}}{D_N} \ , \ n \geqq m \tag{4}$$

where $\mathbf{B}_N = \|b_{mn}\|$ is the inverse of $\mathbf{A}_N$. It remains, therefore, but to calculate D_N.

Let $k = N-1$ in (3). Then

$$D_N = D_{N-1} - \lambda^2 D_{N-2} \ , \ N = 2,3, \cdots \tag{5}$$

may be considered as a linear difference equation with the initial conditions

$$\begin{aligned} D_0 &= 1 \\ D_1 &= 1 \end{aligned} \tag{6}$$

[see Section (e) of Appendix B]. Let $D_N = \mu^N$. Then the roots of the indicial equation

$$\mu^2 - \mu + \lambda^2 = 0 \tag{7}$$

are

$$\mu_{1,2} = \tfrac{1}{2}(1 \pm \sqrt{1 - 4\lambda^2}) \ .$$

If $0 \neq |\lambda| < \frac{1}{2}$, then the roots are real, positive, and unequal. If $|\lambda| = \frac{1}{2}$, the roots are real and equal. If $|\lambda| > \frac{1}{2}$, the roots of the indicial equation are complex. We shall consider all three cases.

Suppose $0 \neq |\lambda| < \frac{1}{2}$. Let μ_1 be the larger and μ_2 the smaller of the roots. Then the solution of the difference system (5)–(6) is

$$D_N = \frac{1}{\mu_1 - \mu_2}(\mu_1^{N+1} - \mu_2^{N+1}) \ , \ N = 0, 1, \cdots \tag{8}$$

(and is always nonzero). Thus from (4)

$$b_{mn} = b_{nm} = \left(\frac{-\lambda}{\mu_1}\right)^{n-m} \frac{1}{\mu_1 - \mu_2}\left[\frac{(1 - \rho^{N-n+1})(1 - \rho^m)}{1 - \rho^{N+1}}\right] \ , \ n \geqq m$$

where

$$\rho = \frac{\mu_2}{\mu_1} < 1 \ .$$

As N increases without limit with m and n fixed,

$$b_{mn} = b_{nm} = \left(\frac{-\lambda}{\mu_1}\right)^{n-m} \frac{(1 - \rho^m)}{\mu_1 - \mu_2} \ , \ n \geqq m$$

and for m large

$$b_{mn} = b_{nm} = \left(\frac{-\lambda}{\mu_1}\right)^{n-m} \frac{1}{\mu_1 - \mu_2} \ , \ n \geqq m \ .$$

We also see from (12.16) that

$$s(n) = \left(\frac{-\lambda}{\mu_1}\right)^{|n|} \frac{1}{\mu_1 - \mu_2}$$

since $\mu_1 = -\lambda z_1$ and $\mu_2 = -\lambda z_2$.

Now let $\lambda = \pm\frac{1}{2}$. Then the solution of the difference system is

$$D_N = (N+1)(\tfrac{1}{2})^N \ , \ N = 0, 1, \cdots$$

and from (4)

$$b_{mn}=b_{nm}=\frac{2(-1)^{n-m}}{N+1}\,m(N+1-n)\ ,\ n\geqq m$$

if $\lambda=+\frac{1}{2}$, while

$$b_{mn}=b_{nm}=\frac{2}{N+1}\,m(N+1-n)\ ,\ n\geqq m$$

if $\lambda=-\frac{1}{2}$. Note that in this latter case

$$\mathbf{B}_n=\frac{2}{n+1}\,\mathbf{S}$$

where $\mathbf{S}$ is the matrix of (7.13).

Finally, if $|\lambda|>\frac{1}{2}$, the roots of the indicial equation (7) are

$$\mu_1=|\lambda|e^{i\theta}$$

$$\mu_2=|\lambda|e^{-i\theta}$$

where

$$0<\theta<\tfrac{1}{2}\pi$$

and

$$\sin\theta=\frac{\sqrt{4\lambda^2-1}}{2|\lambda|}$$

$$\cos\theta=\frac{1}{2|\lambda|}\ .$$

The solution of the difference system (5)–(6) is thus

$$D_N=|\lambda|^N\,\frac{\sin(N+1)\theta}{\sin\theta}\ ,\ N=0,1,\cdots$$

and

$$b_{mn}=b_{nm}=2\left(\frac{-\lambda}{|\lambda|}\right)^{n-m}\frac{\sin(N+1-n)\theta\,\sin m\theta\,\cos\theta}{\sin(N+1)\theta\,\sin\theta}\ ,\ n\geqq m$$

provided $(N+1)\theta$ is not a multiple of π.

(c) Vandermonde matrices

Let t_0, t_1, $\cdots$ be a strictly increasing sequence of positive numbers with $t_0 = 1$. Then

$$\mathbf{V}_n(t_1, \cdots, t_n) = \|t_j^{i-1}\| = \begin{bmatrix} 1 & 1 & 1 & \cdots & 1 \\ t_1 & t_2 & t_3 & \cdots & t_n \\ t_1^2 & t_2^2 & t_3^2 & \cdots & t_n^2 \\ \cdot & \cdot & \cdot & \cdot & \cdot \\ t_1^{n-1} & t_2^{n-1} & t_3^{n-1} & \cdots & t_n^{n-1} \end{bmatrix} \tag{9}$$

is an $n \times n$ *Vandermonde matrix*. Its determinant is

$$\det \mathbf{V}_n(t_1, \cdots, t_n) = \prod_{\substack{i,j=1 \\ i>j}}^{n} (t_i - t_j) = \prod_{i=2}^{n} \left[\prod_{j=1}^{i-1} (t_i - t_j) \right] . \tag{10}$$

The $(n-1) \times (n-1)$ Vandermonde matrix $\mathbf{V}_{n-1}(t_1, \cdots, t_{k-1}, t_{k+1}, \cdots, t_n)$ is identical with the matrix obtained by striking out the nth row and kth column of (9). Its determinant is

$$\det \mathbf{V}_{n-1}(t_1, \cdots, t_{k-1}, t_{k+1}, \cdots, t_n) = [\det \mathbf{V}_n(t_1, \cdots, t_n)] \times (-1)^{n-k} \prod_{\substack{i=1 \\ i \neq k}}^{n} (t_k - t_i)^{-1} .$$

For example, if $t_i = (i+1)^2$, then

$$\det \mathbf{V}_n(t_1, \cdots, t_n) = \frac{2[1!3!5! \cdots (2n+1)!]}{n!(n+1)!(n+2)!}$$

and

$$(-1)^{n-k} \prod_{\substack{i=1 \\ i \neq k}}^{n} (t_k - t_i) = \frac{(n-k)!(n+k+2)!}{2k(k+1)^2(k+2)} .$$

Two closely related matrices are those involving odd and even powers, namely,

$$\mathbf{W}_n(t_1, \cdots, t_{n-1}) = \begin{bmatrix} 1 & 1 & 1 & \cdots & 1 \\ t_1 & t_1^3 & t_1^5 & \cdots & t_1^{2n-1} \\ t_2 & t_2^3 & t_2^5 & \cdots & t_2^{2n-1} \\ \cdot & \cdot & \cdot & \cdot & \cdot \\ t_{n-1} & t_{n-1}^3 & t_{n-1}^5 & \cdots & t_{n-1}^{2n-1} \end{bmatrix} \tag{11}$$

and

$$\mathbf{U}_n(t_1, \cdots, t_{n-1}) = \begin{bmatrix} 1 & 1 & 1 & \cdots & 1 \\ t_1^2 & t_1^4 & t_1^6 & \cdots & t_1^{2n} \\ t_2^2 & t_2^4 & t_2^6 & \cdots & t_2^{2n} \\ \cdot & \cdot & \cdot & \cdot & \cdot \\ t_{n-1}^2 & t_{n-1}^4 & t_{n-1}^6 & \cdots & t_{n-1}^{2n} \end{bmatrix} . \qquad (12)$$

In terms of the Vandermonde matrix we may write

$$\mathbf{W}_n(t_1, \cdots, t_{n-1}) = \|t_{j-1}\delta_{ij}\| \mathbf{V}_n'(t_0^2, \cdots, t_{n-1}^2) \qquad (13)$$

and

$$\mathbf{U}_n(t_1, \cdots, t_{n-1}) = \|t_{j-1}\delta_{ij}\| \mathbf{W}_n(t_1, \cdots, t_{n-1}) \; . \qquad (14)$$

Furthermore, standard manipulation of determinants yields

$$\det \mathbf{W}_n(t_1, \cdots, t_{n-1}) = \left[\prod_{i=1}^{n-1} t_i(t_i^2 - 1) \right] \det \mathbf{V}_{n-1}(t_1^2, \cdots, t_{n-1}^2)$$

and

$$\det \mathbf{U}_n(t_1, \cdots, t_{n-1}) = \left[\prod_{i=1}^{n-1} t_i \right] \det \mathbf{W}_n(t_1, \cdots, t_{n-1}) \; .$$

If, in particular, $t_i = i + 1$, then

$$\det \mathbf{W}_n(t_1, \cdots, t_{n-1}) = 1!3!5! \cdots (2n-1)!$$

and

$$\det \mathbf{U}_n(t_1, \cdots, t_{n-1}) = n![1!3!5! \cdots (2n-1)!] \; .$$

Let σ_m be the elementary symmetric functions of $t_1, t_2, \cdots, t_n$. Let $\sigma_{0,p} = \sigma_0 = 1$ for all p, and let $\sigma_{m,p}$ (for $m \geqq 1$) be σ_m with t_p set equal to zero. If we write

$$\mathbf{V}_n^{-1}(t_1, \cdots, t_n) = \|V_{ij}\| \; ,$$

then

$$V_{ij} = \frac{(-1)^{j-1} \, \sigma_{n-j,i}}{\prod_{\substack{k=1 \\ k \neq i}}^{n} (t_k - t_i)}$$

(see [Macon and Spitzbart]). Thus the inverses of $\mathbf{W}_n$ and $\mathbf{U}_n$ also are determined.

The first row of $\mathbf{W}_n^{-1}$ and $\mathbf{U}_n^{-1}$ frequently is quite useful in certain practical problems. If we let

$$\mathbf{W}_n^{-1}(t_1, \cdots, t_{n-1}) = \|W_{ij}\| \ ,$$

then we may write

$$W_{1k} = \frac{(-1)^{n-1} \prod_{i=1}^{n-1} t_i^2}{t_{k-1}^3 \prod_{\substack{j=0 \\ j \neq k-1}}^{n-1} (t_{k-1}^2 - t_j^2)} \ , \quad k = 1, \cdots, n \ .$$

Of course

$$\sum_{k=1}^{n} W_{1k} t_{k-1} = 1 \ . \tag{15}$$

To verify (15) in the concrete case $t_i = i+1$, we see that

$$W_{1k} = \frac{2(-1)^{k+1}(n!)^2}{k(n-k)!(n+k)!} \ , \quad k = 1, \cdots, n \ ,$$

and (15) becomes

$$2(n!)^2 \sum_{k=1}^{n} \frac{(-1)^{k+1}}{(n-k)!(n+k)!} = 1 \ .$$

But this is just a special case of the formula

$$\sum_{k=0}^{n-m-1} (-1)^k \binom{n}{m+1+k} = \binom{n-1}{m}$$

(see [Riordan, p. 33]).

Now let

$$\mathbf{U}_n^{-1}(t_1, \cdots, t_{n-1}) = \|U_{ij}\| \ .$$

Then

$$U_{1k} = \frac{W_{1k}}{t_{k-1}} \ , \quad k = 1, \cdots, n \ ,$$

and

$$\sum_{k=1}^{n} U_{1k}t_{k-1}^2 = 1 \ .$$

If $t_i = i+1$, then

$$U_{1k} = \frac{2(-1)^{k+1}(n!)^2}{k^2(n-k)!(n+k)!} \ , \ k=1, \cdots, n \ .$$

(d) Geometric matrices

Given a sequence $\{a_n\}$ of numbers we may construct a family of symmetric matrices $\{\mathbf{G}_n\}$ where

$$\mathbf{G}_n = \begin{bmatrix} a_1 & a_2 & \cdots & a_n \\ a_2 & a_3 & \cdots & a_{n+1} \\ \cdot & \cdot & \cdot & \cdot \\ a_n & a_{n+1} & \cdots & a_{2n-1} \end{bmatrix} .$$

We wish to consider the properties of $\mathbf{G}_n$ for various sequences $\{a_n\}$.

Suppose $\{a_n\}$ is a geometric sequence,

$$a_1 = 1$$

$$a_n = \lambda^{n-1} \ , \ n=2,3, \cdots, \quad \lambda \neq 0 \ .$$

Then we call

$$\mathbf{G}_n = \begin{bmatrix} 1 & \lambda & \cdots & \lambda^{n-1} \\ \lambda & \lambda^2 & \cdots & \lambda^n \\ \cdot & \cdot & \cdot & \cdot \\ \lambda^{n-1} & \lambda^n & \cdots & \lambda^{2(n-1)} \end{bmatrix} \tag{16}$$

a *geometric matrix.*

Clearly $\mathbf{G}_n$ has rank one. Hence the only nonzero eigenvalue of $\mathbf{G}_n$ is

$$\operatorname{tr} \mathbf{G}_n = \frac{1-\lambda^{2n}}{1-\lambda^2} \ , \ \lambda \neq \pm 1$$

$$= n \qquad , \ \lambda = \pm 1 \ .$$

[If $\lambda = 1$, then $\mathbf{G}_n = \boldsymbol{\Delta}$, the $n \times n$ matrix with ones everywhere.] We also see from (4.1) that

$$(\mathbf{I}_n + \mathbf{G}_n)^{-1} = \mathbf{I}_n - \frac{1-\lambda^2}{(1-\lambda^2)+(1-\lambda^{2n})} \mathbf{G}_n \ , \quad \lambda \neq \pm 1 \ . \tag{17}$$

(e) Arithmetic matrices

Suppose the sequence $\{a_n\}$ of Section (d) is an arithmetic sequence,

$$a_1 = a$$
$$a_n = a + (n-1)d\ , \quad n = 2,3, \cdots \ .$$

Then we call

$$\mathbf{A}_n = \begin{bmatrix} a & a+d & \cdots & a+(n-1)\mathrm{d} \\ a+d & a+2d & \cdots & a+nd \\ \cdot & \cdot & \cdot & \cdot \\ a+(n-1)d & a+nd & \cdots & a+2(n-1)d \end{bmatrix} \tag{18}$$

an *arithmetic matrix*. It is not difficult to see that $\rho\mathbf{A}_n = 2$ if $d \neq 0$ and that $\rho\mathbf{A}_n = 1$ if $d = 0$ and $a \neq 0$. We shall assume that $n \geqq 2$ and that $d \neq 0$.

Let

$$s_1 = \operatorname{tr} \mathbf{A}_n$$
$$s_2 = \operatorname{tr} \mathbf{A}_n^2 \quad .$$

Since $\mathbf{A}_n$ has rank two, $n-2$ eigenvalues of $\mathbf{A}_n$ are zero. Let μ and ν be the remaining eigenvalues. Then

$$\begin{aligned} \sigma_1 &= s_1 = \mu + \nu \\ \sigma_2 &= \tfrac{1}{2}(s_1^2 - s_2) = \mu\nu \end{aligned} \tag{19}$$

are the elementary symmetric functions of the eigenvalues (see Section 2) and

$$(\mathbf{I}_n + \mathbf{A}_n)^{-1} = \mathbf{I}_n - \frac{1}{1+\sigma_1+\sigma_2}\,[(1+\sigma_1)\mathbf{A}_n - \mathbf{A}_n^2]$$

[see (4.4)]. Thus to find μ, ν and $(\mathbf{I}_n + \mathbf{A}_n)^{-1}$ we need compute merely σ_1 and σ_2.

Trivially

$$\sigma_1 = s_1 = \operatorname{tr} \mathbf{A}_n = na + n(n-1)d \ .$$

Some additional arithmetic also shows that

$$\begin{aligned} s_2 = \operatorname{tr} \mathbf{A}_n^2 &= \sum_{k=0}^{n-1} \sum_{j=k}^{n+k-1} (a+jd)^2 \\ &= n^2a^2 + 2n^2(n-1)ad + \frac{n^2(n-1)(7n-5)}{6}\, d^2 \end{aligned}$$

and hence

$$\sigma_2 = -\frac{n^2(n^2-1)}{12} d^2 . \tag{20}$$

(f) Harmonic matrices

Consider now the sequence $\{a_n^{-1}\}$ where $\{a_n\}$ is the arithmetic sequence of Section (e). Then we call

$$\mathbf{H}_n = \begin{bmatrix} \frac{1}{a} & \frac{1}{a+d} & \cdots & \frac{1}{a+(n-1)d} \\ \frac{1}{a+d} & \frac{1}{a+2d} & \cdots & \frac{1}{a+nd} \\ \cdot & \cdot & \cdot & \cdot \\ \frac{1}{a+(n-1)d} & \frac{1}{a+nd} & \cdots & \frac{1}{a+2(n-1)d} \end{bmatrix} \tag{21}$$

an *harmonic matrix*. (We assume $a+kd \neq 0$ for any integer k.) When $a=1=d$, we call $\mathbf{H}_n$ a *Hilbert matrix*.

If we let

$$\pi_k = \prod_{j=k-1}^{n+k-2} (a+jd) \tag{22}$$

be the product of the denominators in the kth row or column of $\mathbf{H}_n$, then some trivial manipulations show that

$$\det \mathbf{H}_n = \frac{(n-1)!d^{n-1}}{\pi_n} \det \mathbf{K}_n \tag{23}$$

where

$$\mathbf{K}_n = \begin{bmatrix} \frac{1}{a} & \frac{1}{a+d} & \cdots & \frac{1}{a+(n-2)d} & 1 \\ \frac{1}{a+d} & \frac{1}{a+2d} & \cdots & \frac{1}{a+(n-1)d} & 1 \\ \cdot & \cdot & \cdot & \cdot & \cdot \\ \frac{1}{a+(n-1)d} & \frac{1}{a+nd} & \cdots & \frac{1}{a+(2n-3)d} & 1 \end{bmatrix} . \tag{24}$$

Also, see [Muir, pp. 429-431]

$$\det \mathbf{H}_n = \frac{\prod_{k=1}^{n} [(n-k)!d^{n-k}]^2}{\prod_{k=1}^{n} \pi_k} . \tag{25}$$

Let $\mathbf{b}_n' = \{b_1, \cdots, b_n\}$ and $\mathbf{c}_n' = \{c_1, \cdots, c_n\}$ for any n. Let $\mathbf{A}_n$ be a nonsingular symmetric matrix for any n. Then we define

$$\mathbf{A}_{n+1} = \begin{bmatrix} \mathbf{A}_n & \mathbf{b}_n \\ \mathbf{b}_n' & b_{n+1} \end{bmatrix} .$$

and

$$\mathbf{B}_{n+1} = \begin{bmatrix} \mathbf{A}_n & \mathbf{c}_n \\ \mathbf{b}_n' & c_{n+1} \end{bmatrix} .$$

From Section 6 we see that

$$\mathbf{c}_n'\mathbf{A}_n^{-1}\mathbf{c}_n = \mathbf{c}_{n-1}'\mathbf{A}_{n-1}^{-1}\mathbf{c}_{n-1} + \frac{\det \mathbf{A}_{n-1}}{\det \mathbf{A}_n} (c_n - \mathbf{c}_{n-1}'\mathbf{A}_{n-1}^{-1}\mathbf{b}_{n-1})^2$$

and

$$\det \mathbf{B}_n = (c_n - \mathbf{c}_{n-1}'\mathbf{A}_{n-1}^{-1}\mathbf{b}_{n-1}) \det \mathbf{A}_{n-1} .$$

Thus

$$\mathbf{c}_n'\mathbf{A}_n^{-1}\mathbf{c}_n = \mathbf{c}_{n-1}'\mathbf{A}_{n-1}^{-1}\mathbf{c}_{n-1} + \frac{(\det \mathbf{B}_n)^2}{(\det \mathbf{A}_{n-1})(\det \mathbf{A}_n)} . \tag{26}$$

If we let $c_n = 1$ for all n, and let $\mathbf{A}_n = \mathbf{H}_n$, then $\mathbf{B}_n = \mathbf{K}_n$. Now from (25)

$$\frac{\det \mathbf{H}_n}{\det \mathbf{H}_{n-1}} = \frac{1}{a + 2(n-1)d} \left[\frac{(n-1)!d^{n-1}}{\prod_{j=n-1}^{2n-3} (a + jd)} \right]^2$$

and hence from (26)

$$\mathbf{c}_n'\mathbf{H}_n^{-1}\mathbf{c}_n = \mathbf{c}_{n-1}' \; \mathbf{H}_{n-1}^{-1} \; \mathbf{c}_{n-1} + [a + 2(n-1)d] .$$

Inductively we find that

$$\mathbf{c}_n'\mathbf{H}_n^{-1}\mathbf{c}_n = n[a + (n-1)d] .$$

Since $\mathbf{c}_n\mathbf{c}_n' = \mathbf{\Delta}$, we see from (8.12) that

$$\sum_{i,j=1}^{n} H_{ij} = n[a + (n-1)d]$$

where $\mathbf{H}_n^{-1}=\|H_{ij}\|$. Also, from (21) and (25),

$$H_{11}=\frac{a\prod_{j=1}^{n-1}(a+jd)^2}{[(n-1)!d^{n-1}]^2}\ . \tag{27}$$

Thus if $a=1=d$, we have the interesting identity

$$\sum_{i,j=1}^{n} H_{ij}=H_{11}=n^2\ . \tag{28}$$

From the formula [Muir, pp. 348, 322],

$$\det\|(x_i-\alpha_j)^{-1}\|=\frac{(-1)^{\frac{1}{2}n(n-1)}[\det\mathbf{V}_n(x_1,\cdots,x_n)][\det\mathbf{V}_n(\alpha_1,\cdots,\alpha_n)]}{\prod_{i=1}^{n}\left[\prod_{j=1}^{n}(x_i-\alpha_j)\right]}$$

where $\mathbf{V}_n$ is the Vandermonde matrix [see (9)], we easily may show that for $a=1=d$,

$$H_{ij}=\frac{(-1)^{i+j}(n+i-1)!(n+j-1)!}{(i-1)!^2(j-1)!^2(n-i)!(n-j)!(i+j-1)}\ ,\quad 1\leqq i,j\leqq n$$

are the elements of the inverse Hilbert matrix. For example, if $n=4$, then

$$\det\mathbf{H}_4=\frac{1}{6{,}048{,}000}$$

and

$$\mathbf{H}_4^{-1}=\begin{bmatrix} 16 & -120 & 240 & -140 \\ -120 & 1{,}200 & -2{,}700 & 1{,}680 \\ 240 & -2{,}700 & 6{,}480 & -4{,}200 \\ -140 & 1{,}680 & -4{,}200 & 2{,}800 \end{bmatrix}.$$

(g) Gaussian matrices

The Markoffian matrix of Section (a) was based on the covariance function $e^{-|x|}$. We now would like to consider a matrix based on the Gaussian covariance function $e^{-\frac{1}{2}x^2}$. If λ is a positive number unequal to one, we shall call

$$\mathbf{G}_n=\|\lambda^{(i-j)^2}\|=\begin{bmatrix} 1 & \lambda & \lambda^4 & \cdots & \lambda^{(n-1)^2} \\ \lambda & 1 & \lambda & \cdots & \lambda^{(n-2)^2} \\ \cdot & \cdot & \cdot & \cdot & \cdot \\ \lambda^{(n-1)^2} & \lambda^{(n-2)^2} & \lambda^{(n-3)^2} & \cdots & 1 \end{bmatrix} \tag{29}$$

an $n\times n$ *Gaussian matrix.*

It does not seem possible to find $\mathbf{G}_n^{-1}$ in a convenient form for n arbitrary. However, we can find det $\mathbf{G}_n$ quite readily.

Using pivotal condensation with the element in the upper left-hand corner of (29) as the pivot, we may write

$$\det \mathbf{G}_n = \det \mathbf{M}_{n-1}$$

where $\partial \mathbf{M}_{n-1} = n-1$ and

$$\mathbf{M}_{n-1} = \|\lambda^{(\alpha-\beta)^2} - \lambda^{\alpha^2+\beta^2}\|_{\alpha\beta} \ .$$

Now multiply the $(\alpha-1)$ row of $\det \mathbf{M}_{n-1}$ by $\lambda^{2\alpha-1}$ and subtract it from the α row for $\alpha = n-1, n-2, \cdots, 2$. Then we obtain

$$\det \mathbf{G}_n = \det \|\lambda^{(\alpha-\beta)^2}(1-\lambda^{2\beta})\| \ , \quad 1 \leqq \alpha,\beta \leqq n-1$$

$$= (\det \mathbf{G}_{n-1}) \prod_{\beta=1}^{n-1} (1-\lambda^{2\beta}) \ .$$

By induction

$$\det \mathbf{G}_n = \prod_{k=1}^{n-1} (1-\lambda^{2k})^{n-k} \ . \tag{30}$$

(h) Kronecker matrices

Let $\mathfrak{A}$ be the square partitioned matrix

$$\mathfrak{A} = \begin{bmatrix} \mathbf{A} & \mathbf{B} & \cdots & \mathbf{B} \\ \mathbf{B} & \mathbf{A} & \cdots & \mathbf{B} \\ \cdot & \cdot & \cdot & \cdot \\ \mathbf{B} & \mathbf{B} & \cdots & \mathbf{A} \end{bmatrix} = \|\mathbf{A}\delta_{ij} + \mathbf{B}(1-\delta_{ij})\| \tag{31}$$

where $\mathbf{A}$ and $\mathbf{B}$ are square matrices of the same dimension. Thus (31) is a generalization of the example considered in Section 4. If $\mathfrak{A}$ is nonsingular, one approach to finding the inverse of $\mathfrak{A}$ is to write

$$\mathfrak{A}^{-1} = \begin{bmatrix} \mathbf{F} & \mathbf{G} & \cdots & \mathbf{G} \\ \mathbf{G} & \mathbf{F} & \cdots & \mathbf{G} \\ \cdot & \cdot & \cdot & \cdot \\ \mathbf{G} & \mathbf{G} & \cdots & \mathbf{F} \end{bmatrix} = \|\mathbf{F}\delta_{ij} + \mathbf{G}(1-\delta_{ij})\| \ , \tag{32}$$

multiply (31) and (32) together and solve for $\mathbf{F}$ and $\mathbf{G}$. (We have used this technique before in Section 6.) If we equate terms of $\mathfrak{A}\mathfrak{A}^{-1} = \mathbf{I}$, there results

$$\begin{aligned} \mathbf{F} &= \mathbf{A}^{-1} - (n-1)\mathbf{A}^{-1}\mathbf{B}\mathbf{G} \\ &= \{\mathbf{A} - (n-1)\mathbf{B}[\mathbf{A} + (n-2)\mathbf{B}]^{-1}\mathbf{B}\}^{-1} \end{aligned} \tag{33}$$

and

$$\begin{aligned}\mathbf{G} &= [(n-1)\mathbf{B}\mathbf{A}^{-1}\mathbf{B}-\mathbf{A}-(n-2)\mathbf{B}]^{-1}\mathbf{B}\mathbf{A}^{-1} \\ &= -(\mathbf{A}-\mathbf{B})^{-1}\mathbf{A}[\mathbf{A}+(n-1)\mathbf{B}]^{-1}\mathbf{B}\mathbf{A}^{-1}\end{aligned} \tag{34}$$

where $\mathfrak{A}$ consists of $n \times n$ blocks. The second formula of (33) is most easily proved by noting that

$$\{(n-1)[\mathbf{I}-(n-1)\mathbf{C}\mathbf{D}]^{-1}\mathbf{C}-(n-1)\mathbf{C}[\mathbf{I}-(n-1)\mathbf{D}\mathbf{C}]^{-1}\}\mathbf{D}=\mathbf{0}$$

[see (3.1)] and then using (3.2) after letting

$$\mathbf{C}=\mathbf{B}[\mathbf{A}+(n-2)\mathbf{B}]^{-1}$$

$$\mathbf{D}=\mathbf{B}\mathbf{A}^{-1} \ .$$

A simpler and more convenient approach, however, is to observe that $\mathfrak{A}$ may be written as the sum of Kronecker products, namely,

$$\mathfrak{A}=(\mathbf{A}-\mathbf{B}) \otimes \mathbf{I}+\mathbf{B} \otimes \mathbf{\Delta}$$

where $\mathbf{\Delta}$ is the $n \times n$ matrix with ones everywhere. Then by (9.21)

$$\mathfrak{A}^{-1}=(\mathbf{A}-\mathbf{B})^{-1} \otimes \mathbf{I}-[\mathbf{A}+(n-1)\mathbf{B}]^{-1}\mathbf{B}(\mathbf{A}-\mathbf{B})^{-1} \otimes \mathbf{\Delta} \ .$$

From (32) we see that

$$\mathbf{G}=-[\mathbf{A}+(n-1)\mathbf{B}]^{-1}\mathbf{B}(\mathbf{A}-\mathbf{B})^{-1}$$

and

$$\mathbf{F}=(\mathbf{A}-\mathbf{B})^{-1}+\mathbf{G} \ .$$

Comparing these last two formulas with (34) and (33) respectively leads to some matrix identities involving **A** and **B**.

From Theorem 9.10

$$\det\mathfrak{A}=\det(\mathbf{A}-\mathbf{B})^{n-1}\det[\mathbf{A}+(n-1)\mathbf{B}] \ .$$

Also, if $\partial\mathbf{A}=\partial\mathbf{B}=m$, and $\lambda_1, \cdots, \lambda_m$ are the eigenvalues of $\mathbf{A}-\mathbf{B}$ and $\mu_1, \cdots, \mu_m$ the eigenvalues of $\mathbf{A}+(n-1)\mathbf{B}$; then $\lambda_1, \cdots, \lambda_m$ are each eigenvalues of multiplicity $n-1$ of $\mathfrak{A}$, and $\mu_1, \cdots, \mu_m$ are simple eigenvalues of $\mathfrak{A}$ (for a total of mn eigenvalues). See Corollary 9.1.

15. SOME APPLICATIONS OF MATRICES

In this monograph we have presented what we believe to be some interesting properties of matrices. Some of the results have been elegant theoretical relationships, others had a distinctly practical flavor. We hope we have convinced the reader that matrices figure prominently in many diverse problems. In this final section we shall consider a few more eclectic applications. However, let us mention that there are still many more topics related to matrix theory which, for various reasons, we have left untouched.

(a) A bilinear recursion problem

Let y_0 be prescribed, and define y_n recursively by the equation

$$y_{n+1}=\frac{\alpha y_n+\beta}{\gamma y_n+\delta}\ ,\ n=0,1,\cdots \tag{1}$$

where $\alpha\delta-\beta\gamma \neq 0$. Our problem is to find a nonrecursive form for y_n.

Since a bilinear transformation of a bilinear transformation is bilinear, y_n may be written in the form

$$y_n=\frac{a_{n-1}y_0+b_{n-1}}{c_{n-1}y_0+d_{n-1}}\ ,\ n=1,2,\cdots ; \tag{2}$$

and if we substitute (2) into (1) there results

$$y_{n+1}=\frac{(\alpha a_{n-1}+\beta c_{n-1})y_0+(\alpha b_{n-1}+\beta d_{n-1})}{(\gamma a_{n-1}+\delta c_{n-1})y_0+(\gamma b_{n-1}+\delta d_{n-1})}\ . \tag{3}$$

Thus if we let $\mathbf{v}_n'=\{a_n,b_n,c_n,d_n\}$ and $\mathbf{v}_0'=\{\alpha,\beta,\gamma,\delta\}$, then from (2) and (3) we see that

$$\mathbf{v}_n=\mathbf{B}\mathbf{v}_{n-1} \tag{4}$$

where

$$\mathbf{B}=\mathbf{I}_2\otimes\mathbf{M}$$

and

$$\mathbf{M}=\begin{bmatrix}\alpha & \beta\\ \gamma & \delta\end{bmatrix}\ .$$

Iterating (4) we obtain

$$\mathbf{v}_n=\mathbf{B}^n\mathbf{v}_0\ ,\ n=0,1,\cdots \tag{5}$$

—which is our penultimate result.

If $\mathbf{J}$ is the Jordan normal form of $\mathbf{M}$, then there exists a nonsingular matrix $\mathbf{P}$ such that

$$\mathbf{J}=\mathbf{PMP}^{-1} \;,$$

and hence

$$\mathbf{B}^n=\mathbf{I}\otimes\mathbf{P}^{-1}\mathbf{J}^n\mathbf{P} \;. \tag{6}$$

Thus we need determine only $\mathbf{P}$ and the eigenvalues of $\mathbf{M}$ in order to determine y_n. Specifically, if λ and μ are distinct eigenvalues of $\mathbf{M}$, then

$$\mathbf{J}^n=\begin{bmatrix}\lambda^n & 0\\ 0 & \mu^n\end{bmatrix} ,$$

while if $\lambda=\mu$, then

$$\mathbf{J}^n=\begin{bmatrix}\lambda^n & n\epsilon\lambda^{n-1}\\ 0 & \lambda^n\end{bmatrix} ,$$

where ϵ is zero or one.

(b) Gershgorin circles

Let $\mathbf{A}=\|a_{ij}\|$ be an $n\times n$ matrix, λ an eigenvalue of $\mathbf{A}$ and $\mathbf{x}'=\{x_1,\cdots,x_n\}$ the corresponding eigenvector. Then $(\mathbf{A}-\lambda\mathbf{I})\mathbf{x}=\mathbf{0}$ implies that

$$\sum_{j=1}^{n}(a_{ij}-\lambda\delta_{ij})x_j=0$$

for every i. Hence for any i,

$$|a_{ii}-\lambda|\;|x_i|\leqq\sum_{\substack{j=1\\ j\neq i}}^{n}|a_{ij}|\;|x_j| \;. \tag{7}$$

Now let

$$|x_k|=\max_i\,|x_i| \;.$$

Since $\mathbf{x}\neq\mathbf{0}$, we have $|x_k|>0$, and hence from (7) (with $i=k$)

$$|a_{kk}-\lambda|\leqq\frac{1}{|x_k|}\sum_{\substack{j=1\\ j\neq k}}^{n}|a_{kj}|\;|x_j| \;.$$

But $|x_j|/|x_k|\leqq 1$. Hence

$$|a_{kk}-\lambda|\leqq r_k$$

where

$$r_k = \sum_{\substack{j=1 \\ j \neq k}}^{n} |a_{kj}| \ .$$

The circle in the complex λ-plane with radius r_k and center at a_{kk} is called a *Gershgorin circle*, C_k. Hence every eigenvalue of **A** lies in at least one Gershgorin circle.

Let

$$D_\alpha = \bigcup_{j=1}^{n_\alpha} C_{\alpha_j} \ , \ 1 \leqq \alpha \leqq \gamma \leqq n \tag{8}$$

be connected sets where $\Sigma_{\alpha=1}^{\gamma} n_\alpha = n$ and $D_\alpha \cap D_\beta = \emptyset$ for $\alpha \neq \beta$. Then each D_α contains exactly n_α eigenvalues of **A**; for each D_α contains at least n_α eigenvalues of **A**, and no eigenvalue of **A** can be in both D_α and D_β for $\alpha \neq \beta$ since D_α and D_β are disjoint.

(c) Lancaster's formula

Let **x** be a random vector normal $(\mathbf{0},\mathbf{R})$ where **R** is positive definite. Then (see Appendix A)

$$w(\mathbf{x}) = \frac{1}{(2\pi)^{\frac{1}{2}n}(\det \mathbf{R})^{\frac{1}{2}}} e^{-\frac{1}{2}\mathbf{x}'\mathbf{R}^{-1}\mathbf{x}}$$

is the density function of **x**. Of course $w(\mathbf{x})$ is nonnegative and $\int_{-\infty}^{\infty} w(\mathbf{x})d\mathbf{x} = 1$. The moments of the products of positive powers of the components of **x** may be found from the moment generating function. In many applied problems, however, it is necessary to find

$$I_n = \int_{-\infty}^{\infty} (\mathbf{x}'\mathbf{M}\mathbf{x})^n w(\mathbf{x})d\mathbf{x} \ , \ n = 0,1, \cdots \tag{9}$$

where **M** is an arbitrary matrix (see [Don], [Lancaster], [Magnus]). We shall deduce a simple algorithm for computing I_n.

We begin by considering the integral

$$K(\alpha) = \int_{-\infty}^{\infty} w(\mathbf{x})e^{-\frac{1}{2}\alpha(\mathbf{x}'\mathbf{M}\mathbf{x})}d\mathbf{x}$$

where α is a parameter. For $|\alpha|$ sufficiently small, the matrix $\mathbf{R}^{-1}+\alpha\mathbf{M}$ is positive definite (see Theorem 5.1). Hence $K(\alpha)$ exists in a neighborhood of $\alpha=0$ and

$$K(\alpha)=[\det(\mathbf{I}+\alpha\mathbf{MR})]^{-\frac{1}{2}}$$

If we differentiate K, n times with respect to α, and then let $\alpha=0$ we obtain

$$I_n=(-1)^n 2^n \frac{d^n}{d\alpha^n} K(\alpha)\bigg|_{\alpha=0} . \tag{10}$$

Thus the evaluation of I_n reduces to the problem of calculating the derivatives of $\det(\mathbf{I}+\alpha\mathbf{MR})$ with respect to α.

Let $\mathbf{G}(\alpha)=\mathbf{I}+\alpha\mathbf{MR}$. Then from Corollary 5.1

$$\frac{d}{d\alpha}\det\mathbf{G}(\alpha)=[\det\mathbf{G}(\alpha)]a_1(\alpha)$$

$$\frac{d}{d\alpha}a_k(\alpha)=-k\,a_{k+1}(\alpha)\ ,\ k=1,2,\cdots$$

where

$$a_k(\alpha)=\mathrm{tr}\,[\mathbf{MRG}^{-1}(\alpha)]^k\ ,\ k=1,2,\cdots .$$

Thus

$$(-1)^1 2^1\frac{d}{d\alpha}K(\alpha)=K(\alpha)\,a_1(\alpha)$$

$$(-1)^2 2^2\frac{d^2}{d\alpha^2}K(\alpha)=K(\alpha)[a_1^2(\alpha)+2a_2(\alpha)]$$

$$(-1)^3 2^3\frac{d^3}{d\alpha^3}K(\alpha)=K(\alpha)[a_1^3(\alpha)+6a_1(\alpha)a_2(\alpha)+8a_3(\alpha)]$$

etc., and we have a simple algorithm for calculating the derivatives of $K(\alpha)$.

Let $a_k=a_k(0)$. Then

$$a_k=\mathrm{tr}\,(\mathbf{MR})^k\ ,\ k=1,2,\cdots .$$

Since $K(0)=1$ we have the *Lancaster formulas*

$$
\begin{aligned}
I_0 &= 1 \\
I_1 &= a_1 \\
I_2 &= a_1^2+2a_2 \\
I_3 &= a_1^3+6a_1a_2+8a_3 \\
I_4 &= a_1^4+12a_1^2a_2+32a_1a_3+12a_2^2+48a_4 \\
I_5 &= a_1^5+20a_1^3a_2+80a_1^2a_3+60a_1a_2^2+240a_1a_4 \\
&\quad +160a_2a_3+384a_5 \\
I_6 &= a_1^6+30a_1^4a_2+160a_1^3a_3+180a_1^2a_2^2+720a_1^2a_4 \\
&\quad +960a_1a_2a_3+2{,}304a_1a_5+120a_2^3+1{,}440a_2a_4 \\
&\quad +640a_3^2+3{,}840a_6 \\
I_7 &= a_1^7+42a_1^5a_2+280a_1^4a_3+420a_1^3a_2^2+1{,}680a_1^3a_4 \\
&\quad +3{,}360a_1^2a_2a_3+8{,}064a_1^2a_5+840a_1a_2^3+10{,}080a_1a_2a_4 \\
&\quad +4{,}480a_1a_3^2+26{,}880a_1a_6+3{,}360a_2^2a_3+16{,}128a_2a_5 \\
&\quad +13{,}440a_3a_4+46{,}080a_7
\end{aligned}
\tag{11}
$$

etc.

Let $\mathfrak{D}$ be a symbolic derivative operator, and let

$$\mathfrak{D}a_k = ka_{k+1} \ .$$

Then we have the recursive formula

$$I_n = a_1I_{n-1}+2\mathfrak{D}I_{n-1} \ , \ n=1,2,\cdots \tag{12}$$

for computing the I_n. In nonrecursive form we have the symbolic formula

$$I_n = (a_1\mathfrak{I}+2\mathfrak{D})^{n-1}a_1 \tag{13}$$

where $\mathfrak{I}$ is the identity operator.

If the matrix $\mathbf{M}$ in (9) has rank one, then

$$\det \mathbf{G}(\alpha) = 1+\alpha a_1$$

and (10) yields the well-known formula

$$I_n = 1 \cdot 3 \cdot 5 \cdot \cdots \cdot (2n-1)a_1^n$$
$$= \frac{(2n)!}{n!\, 2^n} (\mathrm{tr}\, \mathbf{MR})^n \ . \tag{14}$$

We also may use (9) to calculate the moments of products such as

$$J_n = \int_{-\infty}^{\infty} \left[\prod_{k=1}^{n} (\mathbf{x}'\mathbf{M}_k\mathbf{x}) \right] w(\mathbf{x})d\mathbf{x} \ , \ n = 0, 1, \cdots \tag{15}$$

by the use of the algebraic identities

$$2ab = (a+b)^2 - [a^2 + b^2]$$
$$6abc = (a+b+c)^3 - [(a+b)^3 + (b+c)^3 + (c+a)^3] + [a^3 + b^3 + c^3]$$
$$24abcd = (a+b+c+d)^4 - [(a+b+c)^4 + (b+c+d)^4 + (c+d+a)^4$$
$$+ (d+a+b)^4] + [(a+b)^4 + (b+c)^4 + (c+d)^4 + (d+a)^4$$
$$+ (a+c)^4 + (b+d)^4] - [a^4 + b^4 + c^4 + d^4] \ ,$$

etc. For example,

$$\int_{-\infty}^{\infty} (\mathbf{x}'\mathbf{M}_1\mathbf{x})(\mathbf{x}'\mathbf{M}_2\mathbf{x})(\mathbf{x}'\mathbf{M}_3\mathbf{x})w(\mathbf{x})d\mathbf{x}$$
$$= (\mathrm{tr}\, \mathbf{M}_1\mathbf{R})(\mathrm{tr}\, \mathbf{M}_2\mathbf{R})(\mathrm{tr}\, \mathbf{M}_3\mathbf{R})$$
$$+ 2[(\mathrm{tr}\, \mathbf{M}_1\mathbf{R})(\mathrm{tr}\, \mathbf{M}_2\mathbf{R}\mathbf{M}_3\mathbf{R}) + (\mathrm{tr}\, \mathbf{M}_2\mathbf{R})(\mathrm{tr}\, \mathbf{M}_3\mathbf{R}\mathbf{M}_1\mathbf{R}) \tag{16}$$
$$+ (\mathrm{tr}\, \mathbf{M}_3\mathbf{R})(\mathrm{tr}\, \mathbf{M}_1\mathbf{R}\mathbf{M}_2\mathbf{R})]$$
$$+ 4[\mathrm{tr}\, \mathbf{M}_1\mathbf{R}\mathbf{M}_2\mathbf{R}\mathbf{M}_3\mathbf{R} + \mathrm{tr}\, \mathbf{M}_1\mathbf{R}\mathbf{M}_3\mathbf{R}\mathbf{M}_2\mathbf{R}] \ .$$

(d) Linear least squares

We recall that in the classical theory of linear least squares the basic problem was to determine an estimate of a parameter vector $\mathbf{x}$ based on the vector observation

$$\mathbf{z} = \mathbf{H}\mathbf{x} + \boldsymbol{\nu} \ . \tag{17}$$

If $\mathbf{x}$ is a p-dimensional parameter vector, and $\mathbf{z}$ an n-dimensional observation vector, then $\mathbf{H}$ is an $n \times p$ matrix (of known constants). The random vector $\boldsymbol{\nu}$ is assumed to have mean zero and positive definite covariance matrix, say $\boldsymbol{\Sigma}$. Naturally, many estimators of $\mathbf{x}$ (based on $\mathbf{z}$) may be found.

However, we are interested, generally, in the *Markov estimate* (ME) $\hat{\mathbf{x}}$ of $\mathbf{x}$. We call $\hat{\mathbf{x}}$ a Markov estimate of $\mathbf{x}$ if each component $\hat{x}_k$ of $\hat{\mathbf{x}}$ is a minimum variance linear unbiased estimator of the corresponding component x_k of $\mathbf{x}$. Under the assumption that $n \geqq p$ and that $\mathbf{H}$ has maximal rank p, the familiar arguments show that (see Appendix A)

$$\hat{\mathbf{x}} = (\mathbf{H}'\mathbf{\Sigma}^{-1}\mathbf{H})^{-1}\mathbf{H}'\mathbf{\Sigma}^{-1}\mathbf{z} \tag{18}$$

is the ME of $\mathbf{x}$, and the covariance matrix $\mathbf{P}$ of $\hat{\mathbf{x}}$ is

$$\mathbf{P} = (\mathbf{H}'\mathbf{\Sigma}^{-1}\mathbf{H})^{-1} \ . \tag{19}$$

We shall approach the above problem slightly more generally, and derive some formulas that are useful in various applied problems.

Suppose, then, that

$$\mathbf{z}_k = \mathbf{H}_k\mathbf{x} + \boldsymbol{\nu}_k \ , \quad 1 \leqq k \leqq m \tag{20}$$

are m independent vector observations on the parameter vector $\mathbf{x}$. We assume that $\boldsymbol{\nu}_k$ has mean zero and positive definite covariance matrix $\mathbf{R}_k$. We propose to find a "good" estimate of $\mathbf{x}$ based on the $\mathbf{z}_k$. To do this we must place some restrictions on the $\mathbf{H}_k$. For example, if all the $\mathbf{H}_k$ are zero, then our problem would be meaningless. We therefore shall assume that the partitioned matrix

$$\mathcal{H}' = [\mathbf{H}_1' \cdots \mathbf{H}_m'] \tag{21}$$

has rank p.

Now let us attempt to find matrices $\mathbf{A}_1, \cdots, \mathbf{A}_m$ such that

$$\hat{\mathbf{x}} = \mathbf{A}_1\mathbf{z}_1 + \cdots + \mathbf{A}_m\mathbf{z}_m \tag{22}$$

is an unbiased estimate of $\mathbf{x}$, and such that the trace of the covariance matrix of $\hat{\mathbf{x}}$ is minimized. We shall call such an estimator a *best linear unbiased estimate* (BLUE) of $\mathbf{x}$.

If $\hat{\mathbf{x}}$ is to be unbiased, we must have

$$\mathcal{E}\hat{\mathbf{x}} = \left(\sum_{k=1}^{m} \mathbf{A}_k\mathbf{H}_k\right)\mathbf{x} = \mathbf{x}$$

or

$$\mathbf{I} = \sum_{k=1}^{m} \mathbf{A}_k\mathbf{H}_k \tag{23}$$

where $\mathbf{I}$ is the $p \times p$ identity matrix. Now if the $\mathbf{H}_k$ are arbitrary, it may not be possible to choose matrices $\mathbf{A}_k$ such that (23) holds (suppose all the $\mathbf{H}_k$ were equal and of rank less than p.) However, under our assumption that $\mathcal{H}$ has rank p, it *is* possible to choose matrices $\mathbf{A}_k$ such that (23) holds.

The covariance matrix $\mathbf{P}$ of $\hat{\mathbf{x}}$ is

$$\mathbf{P} = \sum_{k=1}^{m} \mathbf{A}_k \mathbf{R}_k \mathbf{A}_k' , \tag{24}$$

and we must minimize tr $\mathbf{P}$ subject to the condition expressed in (23). Thus we have a constrained minimization problem. Let $\mathbf{\Lambda}$ be a Lagrange matrix multiplier (see Appendix B) and consider

$$w = \operatorname{tr} \sum_{k=1}^{m} \mathbf{A}_k \mathbf{R}_k \mathbf{A}_k' - 2 \operatorname{tr} \mathbf{\Lambda}' \left(\sum_{k=1}^{m} \mathbf{A}_k \mathbf{H}_k \right) .$$

The derivative of the scalar w with respect to the matrix $\mathbf{A}_k$ (see [Miller, 1980]) is

$$\frac{\partial w}{\partial \mathbf{A}_k} = 2\mathbf{A}_k \mathbf{R}_k - 2\mathbf{\Lambda} \mathbf{H}_k' , \quad 1 \leqq k \leqq m .$$

Setting these derivatives equal to the zero matrix leads to

$$\mathbf{A}_k = \mathbf{\Lambda} \mathbf{H}_k' \mathbf{R}_k^{-1} , \tag{25}$$

and from (24)

$$\mathbf{P} = \mathbf{\Lambda} \sum_{k=1}^{m} \mathbf{H}_k' \mathbf{A}_k' .$$

Equation (23) now implies that

$$\mathbf{P} = \mathbf{\Lambda} \tag{26}$$

(and hence $\mathbf{\Lambda}$ is symmetric). From (23) and (25)

$$\mathbf{I} = \mathbf{\Lambda} \sum_{k=1}^{m} \mathbf{H}_k' \mathbf{R}_k^{-1} \mathbf{H}_k$$

and from (26)

$$\mathbf{P}^{-1} = \sum_{k=1}^{m} \mathbf{H}_k' \mathbf{R}_k^{-1} \mathbf{H}_k . \tag{27}$$

Also from (22) and (25),

$$\hat{\mathbf{x}} = \mathbf{P} \sum_{k=1}^{m} \mathbf{H}_k' \mathbf{R}_k^{-1} \mathbf{z}_k . \tag{28}$$

Equations (27) and (28) are our desired formulas. If $m=1$, then (28) becomes identical with the ME given by (18), and (27) becomes identical with (19).

Actually our BLUE $\hat{\mathbf{x}}$ of $\mathbf{x}$ [given by (28)] is slightly more general than it appears. We recall that in order to determine $\hat{\mathbf{x}}$ we minimized the expected value of

$$(\hat{\mathbf{x}}-\mathbf{x})'(\hat{\mathbf{x}}-\mathbf{x}) \tag{29}$$

subject to the constraint $\mathcal{E}\hat{\mathbf{x}}=\mathbf{x}$. Now suppose we determine an unbiased estimate $\hat{\boldsymbol{\xi}}$ of $\mathbf{x}$ by minimizing the expectation of the quadratic form

$$(\hat{\mathbf{x}}-\mathbf{x})'\boldsymbol{\Psi}(\hat{\mathbf{x}}-\mathbf{x}) \tag{30}$$

where $\boldsymbol{\Psi}$ is an arbitrary positive definite matrix. We see that (29) is a special case of (30) when $\boldsymbol{\Psi}$ is the identity matrix. The same arguments as used above show that $\hat{\boldsymbol{\xi}}=\hat{\mathbf{x}}$ with probability one. That is, the estimate is *independent* of our choice of $\boldsymbol{\Psi}$.

Suppose now that we are given a sequence $\{\mathbf{z}_m\}$ of vector observations [see (20)] where

$$\mathbf{z}_m=\mathbf{H}_m\mathbf{x}+\boldsymbol{\nu}_m \ , \quad m=1,2, \cdots \ . \tag{31}$$

Then for each m we may determine a BLUE $\hat{\mathbf{x}}_m$ of $\mathbf{x}$ based on $\mathbf{z}_1, \cdots, \mathbf{z}_m$. We shall show that $\hat{\mathbf{x}}_{m+1}$ depends only on $\hat{\mathbf{x}}_m$ and $\mathbf{z}_{m+1}$. Thus, given $\hat{\mathbf{x}}_m$ we need only $\mathbf{z}_{m+1}$ in order to determine the BLUE of $\mathbf{x}$ based on the *total* data set $\mathbf{z}_1, \cdots, \mathbf{z}_{m+1}$. From the point of view of computer utilization this is a tremendous saving in storage.

From (27) the inverse of the covariance matrix $\mathbf{P}_{m+1}$ of $\hat{\mathbf{x}}_{m+1}$ is

$$\begin{aligned}\mathbf{P}_{m+1}^{-1} &= \sum_{k=1}^{m+1} \mathbf{H}_k'\mathbf{R}_k^{-1}\mathbf{H}_k \\ &= \sum_{k=1}^{m} \mathbf{H}_k'\mathbf{R}_k^{-1}\mathbf{H}_k + \mathbf{H}_{m+1}'\mathbf{R}_{m+1}^{-1}\mathbf{H}_{m+1} \\ &= \mathbf{P}_m^{-1} + \mathbf{H}_{m+1}'\mathbf{R}_{m+1}^{-1}\mathbf{H}_{m+1} \ ,\end{aligned} \tag{32}$$

while from (28)

$$\begin{aligned}\hat{\mathbf{x}}_{m+1} &= \mathbf{P}_{m+1}\sum_{k=1}^{m+1} \mathbf{H}_k'\mathbf{R}_k^{-1}\mathbf{z}_k \\ &= \mathbf{P}_{m+1}[\mathbf{P}_m^{-1}\hat{\mathbf{x}}_m + \mathbf{H}_{m+1}'\mathbf{R}_{m+1}^{-1}\mathbf{z}_{m+1}] \ .\end{aligned} \tag{33}$$

Now substitute $\mathbf{P}_m^{-1}$ as given by (32) into the above equation to obtain

$$\hat{\mathbf{x}}_{m+1} = (\mathbf{I} - \mathbf{G}_{m+1}\mathbf{H}_{m+1})\hat{\mathbf{x}}_m + \mathbf{G}_{m+1}\mathbf{z}_{m+1} \tag{34}$$

where

$$\mathbf{G}_{m+1} = \mathbf{P}_{m+1}\mathbf{H}'_{m+1}\mathbf{R}_{m+1}^{-1} \ . \tag{35}$$

Equations (34) and (32) are recursive forms for the least squares estimator $\hat{\mathbf{x}}_{m+1}$ and its inverse covariance matrix $\mathbf{P}_{m+1}^{-1}$. The matrix $\mathbf{G}_{m+1}$ of (35) is sometimes called the *gain matrix*.

We conclude with a derivation of the BLUE $\hat{\mathbf{x}}$ of the parameter vector $\mathbf{x}$ based on *correlated* observations. The formulas we shall derive for $\hat{\mathbf{x}}$ and its covariance matrix $\mathbf{P}$ [see (39) and (40)], although quite elegant, are less explicit than the formulas [see (28) and (27)] that we obtained in the independent case.

Suppose, then, that as before, the observations $\mathbf{z}_1, \cdots, \mathbf{z}_m$ have means $\mathbf{H}_1\mathbf{x}, \cdots, \mathbf{H}_m\mathbf{x}$ respectively. However, let us assume that the $\mathbf{z}_k$ are no longer necessarily independent. We shall let $\mathcal{R}$ be the covariance matrix of the partitioned vector

$$\mathbf{Z}' = \{\mathbf{z}'_1, \cdots, \mathbf{z}'_m\} \ .$$

Then the block diagonal matrices of $\mathcal{R}$ are precisely the covariance matrices $\mathbf{R}_k$ of the $\mathbf{z}_k$.

As in the independent case we wish to choose matrices $\mathbf{A}_1, \cdots, \mathbf{A}_m$ such that

$$\hat{\mathbf{x}} = \mathbf{A}_1\mathbf{z}_1 + \cdots + \mathbf{A}_m\mathbf{z}_m$$

is a BLUE of $\mathbf{x}$. We may write the above equation succinctly as

$$\hat{\mathbf{x}} = \mathcal{A}\mathbf{Z} \tag{36}$$

where

$$\mathcal{A} = [\mathbf{A}_1 \cdots \mathbf{A}_m]$$

is a partitioned matrix. For $\hat{\mathbf{x}}$ to be unbiased we must have

$$\mathcal{A}\mathcal{H} = \mathbf{I} \tag{37}$$

where $\mathcal{H}'$ is given by (21). The covariance matrix $\mathbf{P}$ of $\hat{\mathbf{x}}$ is given by

$$\mathbf{P} = \mathcal{A}\mathcal{R}\mathcal{A}' \ , \tag{38}$$

and minimizing tr $\mathbf{P}$ subject to the constraint (37) leads, as before, to

$$\hat{\mathbf{x}} = \mathbf{P}\mathcal{H}'\mathcal{R}^{-1}\mathbf{Z} \tag{39}$$

where

$$\mathbf{P} = (\mathcal{H}'\mathcal{R}^{-1}\mathcal{H})^{-1} \quad . \tag{40}$$

[Compare (39) and (40) with the ME formulas of (18) and (19).] If the $\mathbf{z}_1, \cdots, \mathbf{z}_m$ observation vectors are independent, then, of course, (39) and (40) reduce to (28) and the inverse of (27) respectively.

(e) Asymptotic variance of an estimator

Suppose that a certain portion of the flight path of a missile may be represented by a polynomial of the pth degree, say

$$x(t) = x_p + x_{p-1}t + \cdots + x_1 t^{p-1} + x_0 t^p \quad . \tag{41}$$

Let $z(T)$ be a measurement of the vehicle's position at time $t = T$. Of course $z(T)$ will not be exactly $x(T)$; the observation will be corrupted by the addition of noise. Mathematically we may express this fact by writing

$$z(T) = x(T) + \nu(T)$$

where $\nu(T)$ is a mean zero random variable with variance, say σ^2. The size of σ depends in part on the accuracy of the sensor used to measure the position of the vehicle.

Now suppose we make n measurements ($n > p$) of the missile's position, say at times $t = T, 2T, \cdots, nT$, ($T > 0$) and from this information attempt to obtain the best estimate of the position at the latest time, that is, at $t = nT$. We see that this is a least squares problem. In our usual formulation we may write

$$\mathbf{z} = \mathcal{H}_p \mathbf{x} + \boldsymbol{\nu}$$

where $\mathbf{z}' = \{z(T), z(2T), \cdots, z(nT)\}$ is the observation vector, $\mathbf{x}' = \{x_p, x_{p-1}, \cdots, x_1, x_0\}$ is the $(p+1)$-dimensional parameter vector, and $\boldsymbol{\nu}' = \{\nu(T), \nu(2T), \cdots, \nu(nT)\}$ is a mean zero random vector with covariance matrix $\mathcal{E}\boldsymbol{\nu}\boldsymbol{\nu}' = \sigma^2\mathbf{I}$, ($\sigma > 0$). The $n \times (p+1)$ dimensional matrix $\mathcal{H}_p$ of independent variables is

$$\mathcal{H}_p = \begin{bmatrix} 1 & T & T^2 & \cdots & T^p \\ 1 & 2T & (2T)^2 & \cdots & (2T)^p \\ \cdot & \cdot & \cdot & \cdot & \cdot \\ 1 & nT & (nT)^2 & \cdots & (nT)^p \end{bmatrix} . \tag{42}$$

Then the minimum variance estimator $\hat{x}_p$ of $x(nT)$ is

$$\hat{x}_p = \boldsymbol{\gamma}_p'(\mathcal{H}_p'\mathcal{H}_p)^{-1}\mathcal{H}_p'\mathbf{z}$$

where

$$\boldsymbol{\gamma}_p' = \{1, nT, \cdots, (nT)^p\}$$

[note that $x(nT) = \boldsymbol{\gamma}_p'\mathbf{x}$, see (41)] and

$$\operatorname{Var} \hat{x}_p = \sigma^2 \boldsymbol{\gamma}_p'(\mathcal{H}_p'\mathcal{H}_p)^{-1}\boldsymbol{\gamma}_p$$

(see Appendix A).

If $p=0$, then

$$\mathcal{H}_0'\mathcal{H}_0 = n$$

$$\boldsymbol{\gamma}_0' = 1$$

and

$$\operatorname{Var} \hat{x}_0 = \frac{1}{n}\,\sigma^2 \ . \tag{43}$$

If $p=1$, then

$$\mathcal{H}_1'\mathcal{H}_1 = \begin{bmatrix} n & \frac{1}{2}n(n+1)T \\ \frac{1}{2}n(n+1)T & \frac{1}{6}n(n+1)(2n+1)T^2 \end{bmatrix}$$

$$(\mathcal{H}_1'\,\mathcal{H}_1)^{-1} = \frac{2}{(n-1)n(n+1)T^2}\begin{bmatrix} (n+1)(2n+1)T^2 & -3(n+1)T \\ -3(n+1)\mathrm{T} & 6 \end{bmatrix}$$

and

$$\boldsymbol{\gamma}_1' = \{1,\ nT\} \ .$$

Then

$$\operatorname{Var} \hat{x}_1 = \frac{2(2n-1)}{n(n+1)}\,\sigma^2$$

and for n large

$$\operatorname{Var} \hat{x}_1 = \frac{4}{n}\,\sigma^2 \ . \tag{44}$$

Similarly we may show that for $p=2$,

$$\operatorname{Var} \hat{x}_2 = \frac{3(3n^2-3n+2)}{n(n+1)(n+2)}\,\sigma^2$$

and for n large

$$\operatorname{Var} \hat{x}_2 = \frac{9}{n}\,\sigma^2 \ . \tag{45}$$

Now what we would like to do is to show that for a polynomial of arbitrary degree p,

$$\operatorname{Var}\hat{x}_p = \frac{(p+1)^2}{n}\sigma^2 \tag{46}$$

for n large. Certainly (46) is valid for $p=0,1,2$ [see (43), (44), (45)].

To solve this problem we find it convenient to establish a recursion relation (a nonhomogeneous linear first order difference equation) for $\operatorname{Var}\hat{x}_p$.

Without loss of generality let $T=1$, and for convenience let

$$\mathbf{A}_p = \mathcal{H}_p'\mathcal{H}_p \ . \tag{47}$$

Then

$$\operatorname{Var}\hat{x}_p = \sigma^2\boldsymbol{\gamma}_p'\mathbf{A}_p^{-1}\boldsymbol{\gamma}_p$$

and

$$\operatorname{Var}\hat{x}_{p+1} = \sigma^2\boldsymbol{\gamma}_{p+1}'\mathbf{A}_{p+1}^{-1}\boldsymbol{\gamma}_{p+1}$$

where

$$\boldsymbol{\gamma}_{p+1}' = \{\boldsymbol{\gamma}_p', \gamma_{p+1}\} = \{\boldsymbol{\gamma}_p', n^{p+1}\}$$

and

$$\mathbf{A}_{p+1} = \begin{bmatrix} \mathbf{A}_p & \mathbf{b}_p \\ \mathbf{b}_p' & c_p \end{bmatrix}$$

with

$$\mathbf{b}_p' = \sum_{k=1}^{n} k^{p+1}\{1,k,\cdots,k^p\}$$

$$c_p = \sum_{k=1}^{n} k^{2p+2} \ .$$

Now let

$$\mathbf{B}_{p+1} = \begin{bmatrix} \mathbf{A}_p & \boldsymbol{\gamma}_p \\ \mathbf{b}_p' & \gamma_{p+1} \end{bmatrix}$$

which is identical with $\mathbf{A}_{p+1}$ except for the last column. Then from (14.26)

$$\frac{1}{\sigma^2}[\operatorname{Var}\hat{x}_{p+1}-\operatorname{Var}\hat{x}_p]=\boldsymbol{\gamma}'_{p+1}\mathbf{A}_{p+1}^{-1}\boldsymbol{\gamma}_{p+1}-\boldsymbol{\gamma}'_p\mathbf{A}_p^{-1}\boldsymbol{\gamma}_p$$

$$=\frac{(\det\mathbf{B}_{p+1})^2}{(\det\mathbf{A}_p)(\det\mathbf{A}_{p+1})}\ . \tag{48}$$

Thus our problem reduces to one of finding the determinants of the $\mathbf{A}$ and $\mathbf{B}$ matrices. Towards this end let us first compute $\mathbf{A}_p$. From (47) and (42)

$$\mathbf{A}_p=\mathcal{H}'_p\mathcal{H}_p=\begin{bmatrix} S_0 & S_1 & \cdots & S_p \\ S_1 & S_2 & \cdots & S_{p+1} \\ \cdot & \cdot & \cdot & \cdot \\ S_p & S_{p+1} & \cdots & S_{2p}\end{bmatrix}$$

where

$$S_q=\sum_{k=1}^{n}k^q\ .$$

But if $B_m(t)$ are the Bernoulli polynomials, [Miller, 1960]

$$S_q=\frac{1}{q+1}[B_{q+1}(n+1)-B_{q+1}(1)]\ .$$

Since

$$B_m(t)=t^m+\text{lower powers of } t\ ,$$

we see that for n large

$$S_q=\frac{n^{q+1}}{q+1}\ .$$

Thus

$$\det\mathbf{A}_p=n^{(p+1)^2}\det\mathbf{H}_{p+1}$$

where $\mathbf{H}_{p+1}$ is the harmonic matrix of (14.21) (with $a=1=d$). Similarly

$$\det\mathbf{B}_{p+1}=n^{(p+1)(p+2)}\det\mathbf{K}_{p+2}$$

where $\mathbf{K}_{p+2}$ is the matrix of (14.24) (with $a=1=d$).

Thus from (48)

$$\operatorname{Var}\hat{x}_{p+1}-\operatorname{Var}\hat{x}_p=\frac{\sigma^2}{n}\frac{(\det\mathbf{K}_{p+2})^2}{(\det\mathbf{H}_{p+1})(\det\mathbf{H}_{p+2})}$$

and from Section 14

$$\operatorname{Var} \hat{x}_{p+1} = \operatorname{Var} \hat{x}_p + \frac{(2p+3)}{n} \sigma^2 .$$

Inductively, we see the truth of (46). (See [Leskiw and Miller], [Proschan].)

(f) A particular partitioned matrix

In Section 14.d we started with a sequence of numbers $\{a_n\}$ and constructed a family of symmetric matrices,

$$\begin{bmatrix} a_1 & a_2 & \cdots & a_n \\ a_2 & a_3 & \cdots & a_{n+1} \\ \cdot & \cdot & \cdot & \cdot \\ a_n & a_{n+1} & \cdots & a_{2n-1} \end{bmatrix} . \tag{49}$$

For particular sequences $\{a_n\}$ we were led to the geometric matrix $\mathbf{G}_n$ of (14.16) and the arithmetic matrix $\mathbf{A}_n$ of (14.18).

Many trivial and not so trivial extensions of this procedure exist. We shall mention a few. Let

$$\cdots, a_{-1}, a_0, a_1, \cdots$$

be a doubly infinite sequence of numbers and consider the matrix

$$\begin{bmatrix} a_0 & a_1 & \cdots & a_{n-1} \\ a_{-1} & a_0 & \cdots & a_{n-2} \\ \cdot & \cdot & \cdot & \cdot \\ a_{-n+1} & a_{-n+2} & \cdots & a_0 \end{bmatrix} \tag{50}$$

Note that if $a_0 = 0$ and $a_i = -a_{-i}$, $i = 1, 2, \cdots$, then (50) is skewsymmetric. If

$$a_n = \lambda^n \ , \ \lambda \neq 0 \ , \ n = 0, \pm 1, \cdots$$

then

$$\mathbf{G}_n^* = \begin{bmatrix} 1 & \lambda & \cdots & \lambda^{n-1} \\ \lambda^{-1} & 1 & \cdots & \lambda^{n-2} \\ \cdot & \cdot & \cdot & \cdot \\ \lambda^{-n+1} & \lambda^{-n+2} & \cdots & 1 \end{bmatrix}$$

is a matrix of rank one; and if

$$a_n = a + nd \ , \ d \neq 0 \ , \ n = 0, \pm 1, \cdots$$

then

$$\mathbf{A}_n^* = \begin{bmatrix} a & a+d & \cdots & a+(n-1)d \\ a-d & a & \cdots & a+(n-2)d \\ \cdot & \cdot & \cdot & \cdot \\ a-(n-1)d & a-(n-2)d & \cdots & a \end{bmatrix}$$

is a matrix of rank two.

Other variations are possible. For example, the matrices need not be square. Thus

$$\begin{bmatrix} a_1 & a_2 & \cdots & a_n \\ a_2 & a_3 & \cdots & a_{n+1} \\ \cdot & \cdot & \cdot & \cdot \\ a_m & a_{m+1} & \cdots & a_{m+n-1} \end{bmatrix}$$

is an $m \times n$ matrix based on (49); and we may write a similar variant corresponding to (50). Also we may replace the numbers a_n in our sequences by vectors or matrices.

We now shall describe a rather peculiar, yet nevertheless useful, partitioned matrix [see (52)]. Let

$$\mathbf{d}'_\alpha = \{\delta_{\alpha 1}, \cdots, \delta_{\alpha p}\} \ , \quad -\infty < \alpha < \infty$$

be a p-dimensional vector. Then $\mathbf{d}_\alpha$ has a one in the α position if $1 \leqq \alpha \leqq p$, and all the other entries are zero. If $\alpha < 1$ or $\alpha > p$, then $\mathbf{d}_\alpha$ is the zero vector. The inner product of $\mathbf{d}_\alpha$ and $\mathbf{d}_\beta$ is

$$\mathbf{d}'_\alpha \mathbf{d}_\beta = \delta_{\alpha\beta} \ , \ 1 \leqq \alpha, \beta \leqq p \tag{51a}$$

and if one or more of the four inequalities $\alpha < 1$, $\beta < 1$, $\alpha > p$, $\beta > p$ holds, then

$$\mathbf{d}'_\alpha \mathbf{d}_\beta = 0 \ . \tag{51b}$$

Now let

$$\mathbf{H} = \begin{bmatrix} \mathbf{d}_1 & \mathbf{d}_2 & \cdots & \mathbf{d}_{p+m-1} \\ \mathbf{d}_0 & \mathbf{d}_1 & \cdots & \mathbf{d}_{p+m-2} \\ \cdot & \cdot & \cdot & \cdot \\ \mathbf{d}_{-m+2} & \mathbf{d}_{-m+3} & \cdots & \mathbf{d}_p \end{bmatrix} \tag{52}$$

be an $mp \times (m+p-1)$ dimensional partitioned matrix. This matrix occurs in certain problems in the theory of linear hypotheses where we wish to test the hypothesis that a signal does or does not exist in the presence of clutter. In particular we need to find $\mathbf{H}'\mathbf{H}$ and its inverse.

Note that $\mathbf{H}$ may be written in the suggestive form

$$\mathbf{H}=\begin{bmatrix} \mathbf{I}_p & \mathbf{0}_p & \cdots & \mathbf{0}_p \\ \mathbf{0}_p & \mathbf{I}_p & \cdots & \mathbf{0}_p \\ \cdot & \cdot & \cdot & \cdot \\ \mathbf{0}_p & \mathbf{0}_p & \cdots & \mathbf{I}_p \end{bmatrix} \tag{53}$$

where the $\mathbf{0}_p$ are p-dimensional zero column *vectors*. This is *not* the usual way in which we are accustomed to writing partitioned matrices, since $\mathbf{I}_p$ and $\mathbf{0}_p$ do not have the same number of columns (unless p happens to be one).

Let us now calculate $\mathbf{H}'\mathbf{H}$. Write

$$\mathbf{A}=\mathbf{H}'\mathbf{H}=\|a_{ij}\| \ , \ 1 \leqq i,j \leqq m+p-1 \ .$$

Then the element a_{ij} in the ith row and jth column of $\mathbf{A}$ is

$$a_{ij}=\sum_{\gamma=0}^{m-1} \mathbf{d}'_{i-\gamma}\mathbf{d}_{j-\gamma} \tag{54}$$

and from (51) we see that $\mathbf{A}$ is a diagonal matrix. The diagonal elements are

$$a_{ii}=\sum_{\gamma=0}^{m-1} \mathbf{d}'_{i-\gamma}\mathbf{d}_{i-\gamma} \ , \ 1 \leqq i \leqq m+p-1. \tag{55}$$

Now for $\mathbf{d}'_{i-\gamma}\mathbf{d}_{i-\gamma}$ to be unequal to zero we must have $1 \leqq i-\gamma \leqq p$ [see (51)] or

$$i-p \leqq \gamma \leqq i-1 \ .$$

Thus, eliminating the null terms in (55), we have

$$a_{ii}=\sum_{\gamma=\gamma_{0i}}^{\gamma_{li}} \delta_{i-\gamma,i-\gamma}$$

$$=\gamma_{li}-\gamma_{0i}+1$$

where

$$\gamma_{0i}=\max\ (0,\ i-p)$$

$$\gamma_{li}=\min\ (m-1,\ i-1) \ .$$

More explicitly

$$a_{ii}=i \qquad , \qquad 1 \leqq i < \min(m,p)$$

$$a_{ii}=\min(m,p) \ , \quad \min(m,p) \leqq i \leqq \max(m,p)$$

$$a_{ii}=m+p-i \quad , \quad \max(m,p) < i \leqq m+p-1 \ .$$

Since $\mathbf{A}=\mathbf{H}'\mathbf{H}$ is diagonal and nonsingular, the inverse of $\mathbf{A}$ is immediate.

The matrix $\mathbf{H}$ may be visualized geometrically by reference to the graphic representation of Fig. 3, although the analytical representation of (52) is necessary if we wish to perform algebraic operations on $\mathbf{H}$ (as we have just done).

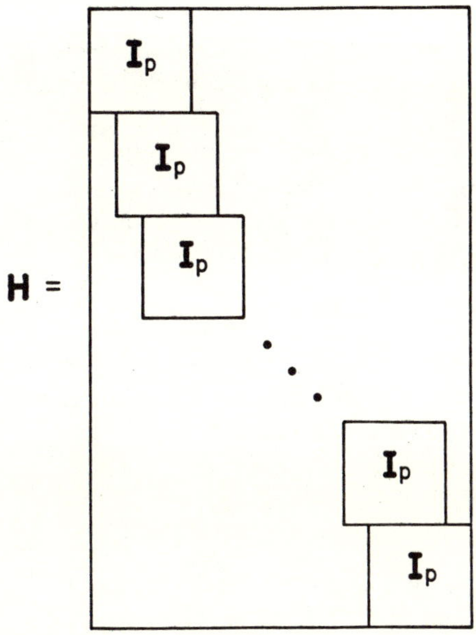

Figure 3

It is not entirely trivial to observe that if the identity matrix $\mathbf{I}_p$ in (53) (or Fig. 3) is replaced by an arbitrary orthogonal matrix $\mathbf{Q}_p$, then the resulting matrix, say $\mathbf{K}$, has the property that

$$\mathbf{K}'\mathbf{K}=\mathbf{H}'\mathbf{H}$$

—independent of $\mathbf{Q}_p$. In fact, analytically,

$$\mathbf{K}=(\mathbf{Q}_p \otimes \mathbf{I}_m)\mathbf{H}$$

and

$$\mathbf{K}'\mathbf{K}=\mathbf{H}'(\mathbf{Q}_p'\otimes\mathbf{I}_m)(\mathbf{Q}_p\otimes\mathbf{I}_m)\mathbf{H}=\mathbf{H}'\mathbf{H} .$$

(g) Linear algebra and vector analysis

In the usual courses in *linear algebra* one considers an n-dimensional space, say $\mathcal{V}_n$, in which certain objects called *vectors* are defined. One also considers linear transformations on these vectors, and at the appropriate point in the development of the theory shows that the linear algebra of all linear transformations from $\mathcal{V}_n$ into $\mathcal{V}_n$ is isomorphic to the linear algebra of all $n\times n$ matrices. Now in the usual courses in *vector analysis*, directed line segments are called *vectors*. One then considers arithmetic operations such as the dot and cross product of two vectors. What is the relationship between these two quantities which are both called vectors? We shall try to answer this question in the present section.

We begin by recalling a few well-known results from linear algebra, and at the same time introduce some convenient notation.

Let $\mathcal{V}_n$ be an n-dimensional unitary space over the field $\mathcal{R}$ of real numbers. That is, $\mathcal{V}_n$ is a linear vector space in which an inner product relationship is defined. If $\mathbf{x}$ and $\mathbf{y}$ are vectors in $\mathcal{V}_n$, we shall denote their inner product by $(\mathbf{x},\mathbf{y})$. The norm $|\mathbf{x}|$ of a vector $\mathbf{x}\in\mathcal{V}_n$ is defined as

$$|\mathbf{x}|=(\mathbf{x},\mathbf{x})^{\frac{1}{2}}\geqq 0 .$$

Of course, $|\mathbf{x}|=0$ if and only if $\mathbf{x}$ is the zero vector. If $|\mathbf{x}|=1$, then we shall call $\mathbf{x}$ a *unit* vector. For convenience we frequently shall write unit vectors as nonbold face characters with a carat, viz.:

$$\hat{x}=\frac{1}{|\mathbf{x}|}\mathbf{x}$$

is a unit vector for all $\mathbf{x}\neq\mathbf{0}$.

Two vectors $\mathbf{x}$ and $\mathbf{y}$ in $\mathcal{V}_n$ are said to be *orthogonal* if $(\mathbf{x},\mathbf{y})=0$. A set α of unit vectors, $\alpha=[\![\hat{e}_{\alpha 1},\cdots,\hat{e}_{\alpha n}]\!]$, is called an *orthonormal basis* in $\mathcal{V}_n$ if $(\hat{e}_{\alpha i},\hat{e}_{\alpha j})=\delta_{ij}$ for $1\leqq i,j\leqq n$. Then any vector $\mathbf{x}\in\mathcal{V}_n$ may be expressed uniquely in the form

$$\mathbf{x}=x_{\alpha 1}\hat{e}_{\alpha 1}+\cdots+x_{\alpha n}\hat{e}_{\alpha n} \tag{56}$$

where $x_{\alpha i}=(\mathbf{x},\hat{e}_{\alpha i})\in\mathcal{R}$, $1\leqq i\leqq n$. If $\beta=[\![\hat{e}_{\beta 1},\cdots,\hat{e}_{\beta n}]\!]$ is also an orthonormal basis in $\mathcal{V}_n$, then $\mathbf{x}$ also may be written as

$$\mathbf{x}=x_{\beta 1}\hat{e}_{\beta 1}+\cdots+x_{\beta n}\hat{e}_{\beta n}$$

where $x_{\beta_i} = (\mathbf{x}, \hat{e}_{\beta_i})$. Thus, given any orthonormal basis γ in $\mathcal{V}_n$, there exists an n-tuple $(x_{\gamma 1}, \cdots, x_{\gamma n})$ which *represents* $\mathbf{x}$ *in that basis.*

Now in matrix algebra we must distinguish between column vectors and row vectors. Let $\mathbf{x} \in \mathcal{V}_n$ be represented in the α-basis by the n-tuple $(x_{\alpha 1}, \cdots, x_{\alpha n})$. Then we write

$$\mathbf{x}_\alpha' = \{x_{\alpha 1}, \cdots, x_{\alpha n}\} \ . \tag{57}$$

It also will be convenient to define

$$(\hat{\mathbf{e}}^\alpha)' = \{\hat{e}_{\alpha 1}, \cdots, \hat{e}_{\alpha n}\} \tag{58}$$

and call $(\hat{\mathbf{e}}^\alpha)'$ a (row) vector. In this notation (56) may be written as

$$\mathbf{x} = \mathbf{x}_\alpha' \hat{\mathbf{e}}^\alpha = (\hat{\mathbf{e}}^\alpha)' \mathbf{x}_\alpha \ . \tag{59}$$

Since $\mathbf{x}$ is independent of the basis chosen, we may refer to $\mathbf{x}$ as a *geometric* (coordinate free) *vector* to distinguish it from the *representation vector* $\mathbf{x}_\alpha$ (in the α-basis) of $\mathbf{x}$, see (57). If, for example, $\mathbf{y} \in \mathcal{V}_n$ is represented by $\mathbf{y}_\alpha' = \{y_{\alpha 1}, \cdots, y_{\alpha n}\}$ in the α-basis, then

$$(\mathbf{x},\mathbf{y}) = (\mathbf{x}_\alpha' \hat{\mathbf{e}}^\alpha, \mathbf{y}_\alpha' \hat{\mathbf{e}}^\alpha) = \mathbf{x}_\alpha' \mathbf{y}_\alpha = \sum_{i=1}^{n} x_{\alpha i} y_{\alpha i} \ . \tag{60}$$

Thus $\mathbf{x}_\alpha' \mathbf{y}_\alpha$ is independent of the basis.

Now let us recall a few elementary facts from the classical theory of vector analysis. Let $n=3$. Then $\mathcal{V}_3$ is three-dimensional Euclidean space. As we have remarked before, a vector $\mathbf{x} \in \mathcal{V}_3$ in this discipline is a directed line segment. Following customary usage we let $\mathbf{i}$, $\mathbf{j}$, $\mathbf{k}$ be unit vectors in a right-handed orthogonal coordinate system, and write $\mathbf{x}$ as

$$\mathbf{x} = x_1 \mathbf{i} + x_2 \mathbf{j} + x_3 \mathbf{k} \tag{61}$$

where x_1, x_2, $x_3 \in \mathcal{R}$. In other words, we have chosen an orthonormal basis $\iota = [\![\mathbf{i},\mathbf{j},\mathbf{k}]\!]$ in $\mathcal{V}_3$ and

$$\mathbf{x}_\iota' = \{x_1, x_2, x_3\}$$

is the representation of $\mathbf{x}$ in the ι-basis.

The dot (or scalar, or inner) product of basis vectors in $\mathcal{V}_3$ is defined as

$$\mathbf{i} \cdot \mathbf{i} = \mathbf{j} \cdot \mathbf{j} = \mathbf{k} \cdot \mathbf{k} = 1$$

$$\mathbf{i} \cdot \mathbf{j} = \mathbf{j} \cdot \mathbf{k} = \mathbf{k} \cdot \mathbf{i} = 0 \ .$$

So the inner product is always a *scalar*. If $\mathbf{x}$ is given by (61) and

$\mathbf{y}=y_1\mathbf{i}+y_2\mathbf{j}+y_3\mathbf{k}$, then

$$\mathbf{x}\cdot\mathbf{y}=x_1y_1+x_2y_2+x_3y_3$$

and, see (60)

$$\mathbf{x}\cdot\mathbf{y}=\mathbf{x}_\iota'\,\mathbf{y}_\iota=(\mathbf{x},\mathbf{y})\ . \tag{62}$$

If $\mathbf{z}=z_1\mathbf{i}+z_2\mathbf{j}+z_3\mathbf{k}$ also is in $\mathcal{V}_3$, then from

$$\mathbf{u}=(\mathbf{x}\cdot\mathbf{y})\mathbf{z}=(x_1y_1+x_2y_2+x_3y_3)z_1\mathbf{i}+(x_1y_1+x_2y_2+x_3y_3)z_2\mathbf{j}$$
$$+(x_1y_1+x_2y_2+x_3y_3)z_3\mathbf{k}$$

follows

$$\mathbf{u}_\iota=\mathbf{P}\mathbf{y}_\iota$$

where

$$\mathbf{P}=\mathbf{z}_\iota\mathbf{x}_\iota'=\begin{bmatrix} z_1x_1 & z_1x_2 & z_1x_3 \\ z_2x_1 & z_2x_2 & z_2x_3 \\ z_3x_1 & z_3x_2 & z_3x_3 \end{bmatrix}\ .$$

Since $(\mathbf{x}\cdot\mathbf{y})\mathbf{z}=(\mathbf{y}\cdot\mathbf{x})\mathbf{z}$ we also may write

$$\mathbf{u}_\iota=\mathbf{Q}\mathbf{x}_\iota$$

where

$$\mathbf{Q}=\mathbf{z}_\iota\mathbf{y}_\iota'=\begin{bmatrix} z_1y_1 & z_1y_2 & z_1y_3 \\ z_2y_1 & z_2y_2 & z_2y_3 \\ z_3y_1 & z_3y_2 & z_3y_3 \end{bmatrix}\ .$$

The cross (or vector, or outer) product of basis vectors is defined as

$$\begin{aligned} \mathbf{i}\times\mathbf{i}&=\mathbf{j}\times\mathbf{j}=\mathbf{k}\times\mathbf{k}=\mathbf{0} \\ \mathbf{i}\times\mathbf{j}&=\mathbf{k}=-\mathbf{j}\times\mathbf{i} \\ \mathbf{j}\times\mathbf{k}&=\mathbf{i}=-\mathbf{k}\times\mathbf{j} \\ \mathbf{k}\times\mathbf{i}&=\mathbf{j}=-\mathbf{i}\times\mathbf{k}\ . \end{aligned} \tag{63}$$

So the outer product is always a vector. If $\mathbf{x}$ and $\mathbf{y}$ are as above, then

$$\mathbf{v}=\mathbf{x}\times\mathbf{y}=(x_2y_3-x_3y_2)\mathbf{i}+(x_3y_1-x_1y_3)\mathbf{j}+(x_1y_2-x_2y_1)\mathbf{k}$$

and

$$\mathbf{v}_\iota=(\mathbf{x}\times\mathbf{y})_\iota=\mathbf{M}\mathbf{y}_\iota$$

where

$$\mathbf{M}=\begin{bmatrix} 0 & -x_3 & x_2 \\ x_3 & 0 & -x_1 \\ -x_2 & x_1 & 0 \end{bmatrix}$$

is a skewsymmetric matrix.

(h) Quaternions

About the turn of the century quaternions were pretty much abandoned in favor of vectors (in the sense of classical vector analysis). In recent years, however, there has been a renewed interest in quaternions. This is due, in part, to the fact that in certain trajectory and orbit analysis problems it is convenient to utilize a four-parameter system in describing phenomena in a three-dimensional space. Thus, in this section, we briefly shall outline some of the simpler properties of quaternions and show their relation to vectors.

Let $\mathcal{Q}$ be the set of all elements ξ of the form

$$\xi=x+\mathbf{x} \tag{64}$$

where x is a scalar, $\iota=[\![\mathbf{i},\mathbf{j},\mathbf{k}]\!]$ an orthonormal basis, $\mathbf{x}=x_1\mathbf{i}+x_2\mathbf{j}+x_3\mathbf{k}$ a vector, and x, x_1, x_2, $x_3 \in \mathcal{R}$, the field of real numbers. Then ξ is called a *quaternion* and $\mathcal{Q}$ is a four-dimensional vector space over the field of real numbers. We define the sum of two quaternions in the expected fashion. And if we treat $\mathbf{i}$, $\mathbf{j}$, $\mathbf{k}$ as indeterminates, the product of two quaternions will be defined as soon as we have defined the multiplication table of the structure constants $\mathbf{i}$, $\mathbf{j}$, $\mathbf{k}$. Our definition of the product of the ι-basis vectors is

$$\begin{aligned} \mathbf{ii}&=\mathbf{jj}=\mathbf{kk}=-1 \\ \mathbf{ij}&=\mathbf{k}=-\mathbf{ji} \\ \mathbf{jk}&=\mathbf{i}=-\mathbf{kj} \\ \mathbf{ki}&=\mathbf{j}=-\mathbf{ik}\ . \end{aligned} \tag{65}$$

Thus if the components are equal, the product is a scalar, while if the components are distinct, the product is a vector. Observe that the rules for multiplying two distinct ι-basis vectors are identical with the outer product rules for vectors [see (63)]. If $\mathbf{x}=x_1\mathbf{i}+x_2\mathbf{j}+x_3\mathbf{k}$ and $\mathbf{y}=y_1\mathbf{i}+y_2\mathbf{j}+y_3\mathbf{k}$, then

$$\mathbf{xy}=\mathbf{x}\times\mathbf{y}-\mathbf{x}\cdot\mathbf{y}\ . \tag{66}$$

With the aid of the above rules we see that the product of ξ [see (64)] and the quaternion $\eta = y + \mathbf{y}$ is

$$\begin{aligned} \xi\eta &= xy + x\mathbf{y} + y\mathbf{x} + \mathbf{x}\mathbf{y} \\ &= xy - \mathbf{x} \cdot \mathbf{y} + x\mathbf{y} + y\mathbf{x} + \mathbf{x} \times \mathbf{y} \\ &= -xy - \mathbf{x} \cdot \mathbf{y} + x\eta + y\xi + \mathbf{x} \times \mathbf{y} \ . \end{aligned} \tag{67}$$

The *conjugate* $\bar{\xi}$ of ξ is defined as

$$\bar{\xi} = x - \mathbf{x} = 2x - \xi$$

and from (67)

$$\xi\bar{\xi} = \bar{\xi}\xi = x^2 + x_1^2 + x_2^2 + x_3^2 = x^2 + |\mathbf{x}|^2 \ .$$

We call

$$N(\xi) = x^2 + x_1^2 + x_2^2 + x_3^2 \tag{68}$$

the *norm* of ξ. If $x = 0$ and $\mathbf{x} = \mathbf{0}$, then we call ξ the zero quaternion. From (68) we see that $\xi = 0$ if and only if $N(\xi) = 0$. Also, (68) implies that every quaternion distinct from the zero quaternion has an inverse, namely,

$$\xi^{-1} = N^{-1}(\xi)\bar{\xi} \ . \tag{69}$$

The conjugate is reflexive, that is $\bar{\bar{\xi}} = \xi$, and some algebra shows that

$$\overline{\xi\eta} = \bar{\eta}\,\bar{\xi} \ . \tag{70}$$

If, in particular, $x = 0$, then $\xi = \mathbf{x}$ is called a *pure* quaternion. The product of two pure quaternions is given by (66).

It immediately follows from (68) that

$$N(\bar{\xi}) = N(\xi) \ ,$$

and if $\xi \neq 0$, eq. (69) implies

$$N(\xi^{-1}) = N^{-1}(\xi) \ .$$

We also see that

$$N(\xi\eta) = N(\xi)N(\eta) \ .$$

To prove this last result we use (70) and the definition of the norm of a quaternion to write

$$N(\xi\eta) = (\xi\eta)(\overline{\xi\eta}) = \xi(\eta\bar{\eta})\bar{\xi} = (\xi\bar{\xi})N(\eta) = N(\xi)N(\eta) \ .$$

Using the above formulas we now may show, that for nonzero quaternions ξ and η,

$$(\xi\eta)^{-1}=\eta^{-1}\xi^{-1} \tag{71}$$

since

$$(\xi\eta)^{-1}=N^{-1}(\xi\eta)\overline{\xi\eta}=N^{-1}(\xi)N^{-1}(\eta)\bar{\eta}\bar{\xi}=\eta^{-1}\xi^{-1} \ .$$

Also, if $\xi \neq 0$,

$$(\xi^{-1})^{-1}=\xi$$

and

$$(\bar{\xi})^{-1}=\overline{\xi^{-1}} \ .$$

The quaternions form a skew field, and considered as a vector space over the field of real numbers, $\mathcal{Q}$ is a noncommutative division algebra. A famous theorem of Frobenius asserts that $\mathcal{Q}$ is the only noncommutative division algebra over $\mathcal{R}$.

We may give a representation of $\mathcal{Q}$ in terms of matrices. Let $\mathcal{M}$ be the vector space of all 2×2 matrices with complex entries, over the field of complex numbers $\mathcal{C}$. Then $\mathcal{M}$ is four-dimensional. It is easy to see that every $\mathbf{M} \in \mathcal{M}$ may be written in the form

$$\mathbf{M}=\begin{bmatrix} a-bi & c-di \\ -c-di & a+bi \end{bmatrix} \tag{72}$$

where a, b, c, d are in $\mathcal{C}$ and i is the imaginary unit. The matrices $\mathbf{I}$, $\mathbf{E}_1$, $\mathbf{E}_2$, $\mathbf{E}_3$ (vectors in $\mathcal{M}$) defined by

$$\begin{aligned} \mathbf{I}&=\begin{bmatrix} 1 & 0 \\ 0 & 1 \end{bmatrix} \\ \mathbf{E}_1&=\begin{bmatrix} -i & 0 \\ 0 & i \end{bmatrix} \\ \mathbf{E}_2&=\begin{bmatrix} 0 & 1 \\ -1 & 0 \end{bmatrix} \\ \mathbf{E}_3&=\begin{bmatrix} 0 & -i \\ -i & 0 \end{bmatrix} \end{aligned} \tag{73}$$

form a basis in $\mathcal{M}$. Let φ be a mapping of $\mathcal{M}$ onto $\mathcal{Q}$ (now considered as a vector space over $\mathcal{C}$) defined by

$$\varphi(\mathbf{M}) = \mu$$

where $\mathbf{M} \in \mathcal{M}$, $\mu \in \mathcal{Q}$ and

$$\varphi(\mathbf{I}) = 1 \ , \ \varphi(\mathbf{E}_1) = \mathbf{i} \ , \ \varphi(\mathbf{E}_2) = \mathbf{j} \ , \ \varphi(\mathbf{E}_3) = \mathbf{k} \ .$$

Since

$$\mathbf{E}_1\mathbf{E}_1 = \mathbf{E}_2\mathbf{E}_2 = \mathbf{E}_3\mathbf{E}_3 = -\mathbf{I}$$

$$\mathbf{E}_1\mathbf{E}_2 = \mathbf{E}_3 = -\mathbf{E}_2\mathbf{E}_1$$

$$\mathbf{E}_2\mathbf{E}_3 = \mathbf{E}_1 = -\mathbf{E}_3\mathbf{E}_2$$

$$\mathbf{E}_3\mathbf{E}_1 = \mathbf{E}_2 = -\mathbf{E}_1\mathbf{E}_3$$

we see that φ is a morphism, and that $\mathcal{M}$ and $\mathcal{Q}$, considered as linear algebras over $\mathcal{C}$, are isomorphic. (Note that $\mathcal{M}$ is not a division algebra since not all nonzero elements of $\mathcal{M}$ have inverses.)

Now the set $\mathcal{N}$ of all 2×2 matrices $\mathbf{N}$ of the form

$$\mathbf{N} = \begin{bmatrix} a - bi & c - di \\ -c - di & a + bi \end{bmatrix}$$

(where $a, b, c, d \in \mathcal{R}$ and i is the imaginary unit) forms a four-dimensional vector space over the field of real numbers. It easily is seen that $\mathcal{N}$ is also a linear algebra over $\mathcal{R}$, and that the matrices $\mathbf{I}$, $\mathbf{E}_1$, $\mathbf{E}_2$, $\mathbf{E}_3$ of (73) form a basis in $\mathcal{N}$. Since

$$\det \mathbf{N} = a^2 + b^2 + c^2 + d^2 \ ,$$

the only singular matrix in $\mathcal{N}$ is the zero matrix. Thus $\mathcal{N}$ is a division algebra, and by Frobenius' theorem, it must be isomorphic to the algebra of quaternions $\mathcal{Q}$ over $\mathcal{R}$—a fact that may be verified directly using arguments employed earlier.

APPENDIX *A*

Stochastic Theory

a. Gaussian Distributions

We recall from elementary probability theory that a random variable X of the continuous type has a probability density function (or frequency function) defined for all X. This frequency function, say p, enjoys the properties that it is nonnegative and that

$$\int_{-\infty}^{\infty} p(X)dX = 1 \quad . \tag{1}$$

The probability that X lie between a and b with $b > a$ is

$$\text{Prob.}[a \leqq X \leqq b] = \int_a^b p(X)dX \quad ,$$

and

$$P(x) = \text{Prob.}[X \leqq x] = \int_{-\infty}^{x} p(X)dX \tag{2}$$

is the distribution function of X. If p is continuous,

$$\frac{d}{dx} P(x) = p(x) \quad .$$

Let f be a deterministic function defined on the real axis, and let the integral of fp exist. Then we write

$$\mathscr{E}f(X) = \int_{-\infty}^{\infty} f(X)p(X)dX$$

and call $\int_{-\infty}^{\infty} f(X)p(X)dX$ "the expected value of f". The symbol $\mathcal{E}$ is called the expectation operator. If

$$\mathcal{E}X^k = \int_{-\infty}^{\infty} X^k p(X)dX \quad , \quad k=0,1, \cdots$$

exists, then $\mathcal{E}X^k$ is the kth moment of X. Clearly

$$\mathcal{E}X^0 = 1 \quad .$$

We call $\mathcal{E}X$ "the mean value of X" while $\mathcal{E}(X-\mathcal{E}X)^2$ is called "the variance of X". The variance is written as

$$\mathrm{Var}X = \mathcal{E}(X-\mathcal{E}X)^2 = \int_{-\infty}^{\infty} (X-\mathcal{E}X)^2 p(X)dX \quad .$$

The function ϕ,

$$\phi(S) = \mathcal{E}e^{SX} = \int_{-\infty}^{\infty} e^{SX} p(X)dX$$

(if it exists) is called the moment generating function. From it we may determine by differentiation the integral moments of X. For example,

$$\phi'(S) = \int_{-\infty}^{\infty} Xe^{SX} p(X)dX \quad ,$$

and letting $S=0$ we obtain

$$\phi'(0) = \int_{-\infty}^{\infty} Xp(X)dX = \mathcal{E}X \quad .$$

Similarly

$$\mathcal{E}X^k = \phi^{(k)}(0) \quad .$$

Knowing the various integral moments of a random variable gives us considerable information concerning the random variable (although it does not necessarily uniquely determine the density function, [Cramér], [Miller, 1980].)

The best known and most useful density function from both a practical and theoretical viewpoint is the *normal* (or *Gaussian*) density function. It is defined for all x by the equation

$$p(x) = \frac{1}{\sigma\sqrt{2\pi}} e^{-(x-\mu)^2/2\sigma^2} \quad , \quad \sigma > 0$$

where μ and σ are parameters. Clearly $p(x) > 0$ for all x, and an elementary integration shows that p has area one. For this normal distribution the mean is

$$\mathcal{E}x = \mu$$

and the variance is

$$\text{Var } x = \mathcal{E}(x-\mu)^2 = \sigma^2 \ .$$

We sometimes say "x is normal (μ,σ^2)". The parameters μ and σ, as well as having physical significance, completely specify the normal distribution. The moment generating function is

$$\phi(s) = \mathcal{E}e^{sx} = \frac{1}{\sigma\sqrt{2\pi}}\int_{-\infty}^{\infty} e^{sx}e^{-(x-\mu)^2/2\sigma^2}dx = e^{\frac{1}{2}\sigma^2 s^2 + \mu s} \ .$$

Now suppose we have a *pair* of random variables, say X and Y. Then they have a joint density function, say $p(X,Y)$. Of course we must have $p(X,Y) \geqq 0$ for all X and Y, and

$$\int_{-\infty}^{\infty}\int_{-\infty}^{\infty} p(X,Y)dXdY = 1 \ .$$

The probability that (X,Y) lie in the rectangle $[a,b]\times[c,d]$ is

$$\text{Prob.}[a \leqq X \leqq b,\ c \leqq Y \leqq d] = \int_a^b dX \int_c^d dY\ p(X,Y)$$

and the corresponding joint distribution function is

$$P(x,y) = \text{Prob.}[X \leqq x,\ Y \leqq y] = \int_{-\infty}^{x} dX \int_{-\infty}^{y} dY\ p(X,Y) \ .$$

Thus, assuming the joint continuity of p,

$$p(x,y) = \frac{\partial^2}{\partial x \partial y} P(x,y) \ .$$

The marginal distribution $P(x)$ of X is given by

$$P(x) = \text{Prob.}[X \leqq x,\ Y < \infty] = \text{Prob.}[X \leqq x]$$

$$= P(x,\infty) = \int_{-\infty}^{\infty} dY \int_{-\infty}^{x} dX\ p(X,Y) = \int_{-\infty}^{x} p(X)dX$$

where $p(X)$ is the frequency function of X. A similar definition holds for the marginal distribution $Q(Y)$ of Y. The marginal density function of X is $p(X)$,

$$p(x)=\frac{\partial}{\partial x}P(x,\infty)=\int_{-\infty}^{\infty}p(x,Y)dY \ ,$$

with a similar definition for the marginal density function $q(Y)$ of Y.

The *conditional* distribution function of X relative to the hypothesis that $Y=y$ is

$$\text{Prob.}[X \leqq x|y]=P(x|y)=\lim_{\substack{h\to 0\\ h>0}}\frac{\text{Prob.}[X \leqq x, y \leqq Y \leqq y+h]}{\text{Prob.}[y \leqq Y \leqq y+h]}$$

$$=\lim_{\substack{h\to 0\\ h>0}}\frac{\int_{-\infty}^{x}dX\int_{y}^{y+h}dY\,p(X,Y)}{\int_{y}^{y+h}q(Y)dY}$$

$$=\frac{\int_{-\infty}^{x}p(X,y)dX}{q(y)}$$

while the corresponding conditional frequency function $p(X|Y)$ is

$$p(x|y)=\frac{\partial}{\partial x}P(x|y)=\frac{p(x,y)}{q(y)} \ .$$

Similar expressions, of course, exist for $Q(y|x)$ and $q(Y|X)$ where these functions are defined in the expected fashion.

If $f(X,Y)$ is a deterministic function of X and Y defined on the whole plane, then we write

$$\mathcal{E}f(X,Y)=\int_{-\infty}^{\infty}\int_{-\infty}^{\infty}f(X,Y)p(X,Y)dXdY$$

(provided the integral exists). If

$$\mathcal{E}X=\int_{-\infty}^{\infty}\int_{-\infty}^{\infty}Xp(X,Y)dXdY=\int_{-\infty}^{\infty}Xp(X)dX$$

exists, we call it the mean of X. Similarly for $\mathcal{E}Y$. A particularly important cross moment (if it exists) is the *covariance* of X and Y defined as

$$\mathcal{E}(X-\mathcal{E}X)(Y-\mathcal{E}Y) \ .$$

It is a measure of the correlation between X and Y.

We also may define the conditional expectation $\mathcal{E}(X|Y)$ (provided it exists) as

$$\mathcal{E}(X|Y) = \int_{-\infty}^{\infty} Xp(X|Y)dX \ .$$

In general it will be a function of Y. Similarly we define $\mathcal{E}(Y|X)$.

Now suppose $\mathbf{X}$ is a random *vector*. That is, $\mathbf{X}' = \{X_1, \cdots, X_n\}$ where $X_1, \cdots, X_n$ are random variables. Then $\mathbf{X}$ has an n-dimensional density function, say $p(\mathbf{X})$, and the n-dimensional integral

$$\mathcal{E}f(\mathbf{X}) = \int_{-\infty}^{\infty} f(\mathbf{X})p(\mathbf{X})d\mathbf{X}$$

over all of n-dimensional Euclidean space (provided it exists) is the expected value of the deterministic function $f(\mathbf{X})$. We call

$$\mathcal{E}X_1^{k_1}X_2^{k_2} \cdots X_n^{k_n} = \int_{-\infty}^{\infty} X_1^{k_1} X_2^{k_2} \cdots X_n^{k_n} \, p(\mathbf{X})d\mathbf{X}$$

(where $k_1, k_2, \cdots, k_n$ are nonnegative integers) the mixed integral moments of $\mathbf{X}$. In particular, if $\mathcal{E}X_i = \mu_i$, $1 \leqq i \leqq n$, then

$$\mathcal{E}\mathbf{X} = \boldsymbol{\mu}$$

where $\boldsymbol{\mu}' = \{\mu_1, \cdots, \mu_n\}$. We call $\boldsymbol{\mu}$ the *mean vector*. The quantity

$$m_{ij} = \mathcal{E}(X_i - \mu_i)(X_j - \mu_j) \ , \ 1 \leqq i,j \leqq n \ ,$$

is called the *covariance* of X_i and X_j. Thus if $i=j$, m_{ii} is the variance of X_i.

The matrix

$$\mathbf{M} = \|m_{ij}\| = \mathcal{E}(\mathbf{X} - \boldsymbol{\mu})(\mathbf{X} - \boldsymbol{\mu})'$$

is the *covariance matrix* of $\mathbf{X}$. A very important property of the covariance matrix is that it is nonnegative definite. This crucial property is easy to prove. For, let $\boldsymbol{\lambda}$ be any n-dimensional vector. Then

$$\mathcal{E}[\boldsymbol{\lambda}'(\mathbf{X} - \boldsymbol{\mu})]^2 \geqq 0 \ . \tag{3}$$

But

$$[\boldsymbol{\lambda}'(\mathbf{X} - \boldsymbol{\mu})]^2 = \boldsymbol{\lambda}'(\mathbf{X} - \boldsymbol{\mu})(\mathbf{X} - \boldsymbol{\mu})'\boldsymbol{\lambda}$$

and

$$\mathcal{E}\boldsymbol{\lambda}'(\mathbf{X} - \boldsymbol{\mu})(\mathbf{X} - \boldsymbol{\mu})'\boldsymbol{\lambda} = \boldsymbol{\lambda}'[\mathcal{E}(\mathbf{X} - \boldsymbol{\mu})(\mathbf{X} - \boldsymbol{\mu})']\boldsymbol{\lambda} = \boldsymbol{\lambda}'\mathbf{M}\boldsymbol{\lambda} \ .$$

By (3), $\boldsymbol{\lambda}'\mathbf{M}\boldsymbol{\lambda} \geqq 0$. That is, $\mathbf{M}$ is nonnegative definite. Generally we assume $\mathbf{M}$ to be positive definite so that det $\mathbf{M} > 0$.

The moment generating function $\phi(\mathbf{S})$ (if it exists) is

$$\phi(\mathbf{S}) = \mathcal{E}e^{\mathbf{S}'\mathbf{X}} = \int_{-\infty}^{\infty} e^{\mathbf{S}'\mathbf{X}} p(\mathbf{X})\, d\mathbf{X}$$

and

$$\frac{\partial}{\partial \mathbf{S}}\, \phi(\mathbf{S}) = \int_{-\infty}^{\infty} \mathbf{X} p(\mathbf{X})\, e^{\mathbf{S}'\mathbf{X}} d\mathbf{X}$$

so that

$$\frac{\partial}{\partial \mathbf{S}} \phi(\mathbf{S}) \bigg|_{\mathbf{S}=\mathbf{0}} = \int_{-\infty}^{\infty} \mathbf{X} p(\mathbf{X}) d\mathbf{X} = \mathcal{E}\mathbf{X} = \boldsymbol{\mu}\ . \tag{4}$$

Similarly

$$\frac{\partial^2}{\partial \mathbf{S}' \partial \mathbf{S}}\ \phi(\mathbf{S}) \bigg|_{\mathbf{S}=\mathbf{0}} = \int_{-\infty}^{\infty} \mathbf{X}\mathbf{X}'\ p(\mathbf{X}) d\mathbf{X} = \mathcal{E}\mathbf{X}\mathbf{X}'\ . \tag{5}$$

Thus the covariance matrix

$$\begin{aligned} \mathbf{M} &= \mathcal{E}(\mathbf{X} - \boldsymbol{\mu})(\mathbf{X} - \boldsymbol{\mu})' \\ &= \mathcal{E}\mathbf{X}\mathbf{X}' - (\mathcal{E}\mathbf{X})(\mathcal{E}\mathbf{X})' \end{aligned}$$

may be deduced from (4) and (5).

If $\mathbf{M}$ is positive definite and

$$p(\mathbf{X}) = \frac{1}{(2\pi)^{\frac{1}{2}n} (\det \mathbf{M})^{\frac{1}{2}}}\ e^{-\frac{1}{2}(\mathbf{X} - \boldsymbol{\mu})'\mathbf{M}^{-1}(\mathbf{X} - \boldsymbol{\mu})} \tag{6}$$

then we say that $\mathbf{X}$ is normally distributed $(\boldsymbol{\mu}, \mathbf{M})$ in n dimensions. Clearly $p(\mathbf{X}) > 0$ for all $\mathbf{X}$ and

$$\int_{-\infty}^{\infty} p(\mathbf{X}) d\mathbf{X} = 1$$

$$\mathcal{E}\mathbf{X} = \boldsymbol{\mu} \tag{7}$$

$$\text{Var}\ \mathbf{X} = \mathcal{E}(\mathbf{X} - \boldsymbol{\mu})(\mathbf{X} - \boldsymbol{\mu})' = \mathbf{M}\ . \tag{8}$$

Let us prove these results. Since $\mathbf{M}$ is positive definite, there exists a nonsingular matrix $\mathbf{R}$ such that $\mathbf{M}^{-1} = \mathbf{R}'\mathbf{R}$. Thus we may write

$$p(\mathbf{X}) = \frac{|\det \mathbf{R}|}{(2\pi)^{\frac{1}{2}n}} e^{-\frac{1}{2}|\mathbf{R}(\mathbf{X}-\boldsymbol{\mu})|^2} .$$

Now make the change of variable $\mathbf{Y} = \mathbf{R}(\mathbf{X} - \boldsymbol{\mu})$ with Jacobian $|\det \mathbf{R}|$. Then

$$\int_{-\infty}^{\infty} p(\mathbf{X}) d\mathbf{X} = \left[\frac{1}{\sqrt{2\pi}} \int_{-\infty}^{\infty} e^{-\frac{1}{2}y^2} \, dy \right]^n = 1 .$$

The moment generating function is

$$\phi(\mathbf{S}) = \mathcal{E} e^{\mathbf{S}'\mathbf{X}} = \frac{1}{(2\pi)^{\frac{1}{2}n} (\det \mathbf{M})^{\frac{1}{2}}} \int_{-\infty}^{\infty} e^{\mathbf{S}'\mathbf{X}} e^{-\frac{1}{2}(\mathbf{X}-\boldsymbol{\mu})'\mathbf{M}^{-1}(\mathbf{X}-\boldsymbol{\mu})} d\mathbf{X} .$$

If we make the same change of variable as above, then

$$\phi(\mathbf{S}) = \frac{1}{(2\pi)^{\frac{1}{2}n}} e^{\boldsymbol{\mu}'\mathbf{S}} \int_{-\infty}^{\infty} e^{\mathbf{S}'\mathbf{R}^{-1}\mathbf{Y}} e^{-\frac{1}{2}\mathbf{Y}'\mathbf{Y}} d\mathbf{Y} .$$

Now we may complete the square in the exponent of the integrand to write

$$\mathbf{Y}'\mathbf{Y} - 2\mathbf{S}'\mathbf{R}^{-1}\mathbf{Y} = (\mathbf{Y} - \mathbf{V})'(\mathbf{Y} - \mathbf{V}) - \mathbf{V}'\mathbf{V}$$

where $\mathbf{V}' = \mathbf{S}'\mathbf{R}^{-1}$. Thus

$$\mathbf{V}'\mathbf{V} = \mathbf{S}'(\mathbf{R}'\mathbf{R})^{-1}\mathbf{S} = \mathbf{S}'\mathbf{M}\mathbf{S}$$

and

$$\phi(\mathbf{S}) = \mathcal{E} e^{\mathbf{S}'\mathbf{X}} = e^{\frac{1}{2}\mathbf{S}'\mathbf{M}\mathbf{S} + \boldsymbol{\mu}'\mathbf{S}} .$$

Applying (4) and (5) to $\phi(\mathbf{S})$ then yields (7) and (8).

b. Stochastic Processes

A stochastic process is a family of random variables [Doob]. If T is a linear index set, then we may write a stochastic process as

$$\mathcal{X} = \{x(t) | \, t \in T\} . \tag{1}$$

That is, $x(t_1)$ is a random variable for every $t_1 \in T$. The most common situations arise when T is an interval (finite or infinite) or a discrete set of points. For example,

$$\mathcal{X} = \{x(t) | -\infty < t < \infty\} \tag{2}$$

is a stochastic process if for every real number t_1, $x(t_1)$ is a random variable.

If x_t is a random variable for every integer t, then the stochastic process

$$\{x_t | t = 0, \pm 1, \cdots\} \tag{3}$$

is generally called a *time series* [see Section (c)].

For any $x(t) \in \mathcal{X}$, the mean of $x(t)$ is $\mathcal{E}x(t)$ and if $\mathcal{E}[x(t) - \mathcal{E}x(t)]^2 < \infty$ we call

$$R(t,s) = \mathcal{E}[x(t) - \mathcal{E}x(t)][x(s) - \mathcal{E}x(s)]$$

the *covariance function* of $\mathcal{X}$. If $\mathcal{E}x(t) = 0$ for all $t \in T$, then we call $\mathcal{X}$ a *mean zero process*. For a mean zero process the covariance function is thus

$$R(t,s) = \mathcal{E}x(t)x(s) \ .$$

Now suppose $\mathcal{E}x(t) = \mu$ (independent of t) for all $t \in T$ and

$$R(t,s) = \mathcal{E}[x(t) - \mu][x(s) - \mu]$$

depends only on the difference of its arguments:

$$R(t,s) \equiv R(t-s) \ .$$

Then we call $\mathcal{X}$ a (wide-sense) *stationary process*.

Various special stochastic processes play distinguished roles in both theory and applications. We mention a few. Suppose T is an interval, as in (1). Then

(i) If $\mathcal{E}[x(t) - x(s)]^2 < \infty$ for all $t,s \in T$ and

$$\mathcal{E}[x(t_2) - x(t_1)][x(t_4) - x(t_3)] = \mathcal{E}[x(t_2) - x(t_1)]\mathcal{E}[x(t_4) - x(t_3)]$$

for any nonoverlapping intervals $[t_1,t_2]$ and $[t_3,t_4]$ in T, then we shall say that $\mathcal{X}$ is a *process with uncorrelated increments*.

(ii) If for all positive integers p and all sets of p distinct points $t_1, \cdots, t_p$ in T the vector $\mathbf{x}' = \{x(t_1), \cdots, x(t_p)\}$ has a joint p-dimensional normal distribution [see (a.6)] then we say $\mathcal{X}$ is a *Gaussian process*.

(iii) If $\mathcal{B} = \{x(t) | t \geqq 0\}$ is a mean zero Gaussian process and Prob.$[x(0) = 0] = 1$, then if

$$\mathcal{E}x(t)x(s) = \sigma^2 \min(t,s) \ , \quad \sigma > 0 \ , \tag{4}$$

we call $\mathcal{B}$ a *Brownian motion process* (sometimes also called a Wiener or Wiener-Lévy process).

(iv) If for every positive integer p and distinct points $t_1, \cdots, t_p$ in T with $t_1 < t_2 < \cdots < t_p$ we have

$$\text{Prob.}[x(t_p) \leqq \xi | x(t_{p-1}), \cdots, x(t_1)] = \text{Prob.}[x(t_p) \leqq \xi | x(t_{p-1})]$$

then $\mathcal{X}$ is said to be a *Markov process.*

We indicate some relations among the above processes. Suppose $\mathcal{B}$ is a Brownian motion process. If $[s,t]$ and $[u,v]$ are two nonoverlapping intervals in T, then

$$\mathcal{E}[x(t)-x(s)][x(v)-x(u)]=0 \ .$$

Hence $\mathcal{B}$ is a process with uncorrelated increments. From (4) we may easily show that

$$\mathcal{E}[x(t)-x(s)]^2 = \sigma^2 |t-s| \ . \tag{5}$$

Now suppose $\mathcal{X}$ [see (2)] is a mean zero Gaussian process which is also a Markov process. Let $R(t,s) = \mathcal{E}x(t)x(s)$ be the covariance function of $\mathcal{X}$. Then if $t_1 < t_2 < t_3$, the covariance matrix $\mathbf{R}$ of $\mathbf{x}' = \{x(t_1), x(t_2), x(t_3)\}$ is

$$\mathbf{R} = \|R_{ij}\| = \|\mathcal{E}x(t_i)x(t_j)\|$$

where $R_{ij} = R(t_i,t_j) = \mathcal{E}x(t_i)x(t_j)$. We shall assume that the 3×3 dimensional matrix $\mathbf{R}$ is positive definite. The conditional expectation of $x(t_3)$ given $x(t_1)$ and $x(t_2)$ is

$$\mathcal{E}(x(t_3)|x(t_2),x(t_1)) = -\frac{\begin{vmatrix} R_{12} & R_{22} \\ R_{13} & R_{23} \end{vmatrix}}{\begin{vmatrix} R_{11} & R_{12} \\ R_{12} & R_{22} \end{vmatrix}} x(t_1) + \frac{\begin{vmatrix} R_{11} & R_{12} \\ R_{13} & R_{23} \end{vmatrix}}{\begin{vmatrix} R_{11} & R_{12} \\ R_{12} & R_{22} \end{vmatrix}} x(t_2) \tag{6}$$

and the conditional expectation of $x(t_3)$ given $x(t_2)$ is

$$\mathcal{E}(x(t_3)|x(t_2)) = \frac{R_{23}}{R_{22}} x(t_2) \ . \tag{7}$$

Since $\mathcal{X}$ is also a Markov process, we must have

$$\mathcal{E}(x(t_3)|x(t_2),x(t_1)) = \mathcal{E}(x(t_3)|x(t_2))$$

and from (6) and (7) we infer that

$$\begin{vmatrix} R_{12} & R_{22} \\ R_{13} & R_{23} \end{vmatrix} = 0$$

or

$$R(t_3,t_2)R(t_2,t_1)=R(t_3,t_1)R(t_2,t_2) \quad . \tag{8}$$

Now if in addition to the Gaussian and Markoffian assumptions we make the additional assumption that $\mathcal{X}$ is stationary, then we may write (8) as

$$R(t_3-t_2)R(t_2-t_1)=R(t_3-t_1)R(0)$$

or

$$R(\lambda)R(\mu)=R(0)R(\lambda+\mu) \tag{9}$$

where $\lambda=t_2-t_1>0$ and $\mu=t_3-t_2>0$. It may be shown [Feller, p. 459] that the only nonidentically zero bounded solution of (9) is

$$R(\nu)=R(0)e^{-\alpha\nu} \quad , \quad \nu>0$$

where α is a nonnegative number. Note that for this $R(\nu)$,

$$\frac{R_{23}}{R_{22}}=\frac{\begin{vmatrix} R_{11} & R_{12} \\ R_{13} & R_{23} \end{vmatrix}}{\begin{vmatrix} R_{11} & R_{12} \\ R_{12} & R_{22} \end{vmatrix}}$$

[see (6) and (7)]. Thus we conclude that the covariance function of a real mean zero stationary Gaussian Markov process is exponential.

c. Time Series

Let $\mathcal{X}=\{x_t|t=0,\pm 1,\cdots\}$ be a mean zero stationary time series with covariance function

$$R(h)=\mathcal{E}x(t+h)x(t) \quad ,$$

[Anderson, 1971]. With the covariance sequence $\{R(h)|h=0,\pm 1,\cdots\}$ we may associate the Fourier series

$$W(\lambda)\sim\frac{1}{2\pi}\sum_{h=-\infty}^{\infty} R(h)e^{-i\lambda h} \quad . \tag{1}$$

If the series on the right-hand side of (1) converges, then we call $W(\lambda)$ the *spectral density* function of $\mathcal{X}$. The spectral density is an even function.

A simple sufficient condition to insure the convergence of the series is to require that the series $\Sigma_{h=0}^{\infty} R(h)$ converge absolutely. Then by the Weierstrass M-test (see Appendix B), the right-hand side of (1) converges absolutely and uniformly in any interval of the λ-axis.

In this case

$$W(\lambda)=\frac{1}{2\pi}\sum_{h=-\infty}^{\infty} R(h)e^{-i\lambda h} \tag{2}$$

represents a continuous function, and we may multiply (2) by $e^{it\lambda}$ and integrate from $-\pi$ to π to obtain

$$R(t)=\int_{-\pi}^{\pi} W(\lambda)e^{it\lambda}d\lambda \ , \ t=0,\pm 1,\cdots \ .$$

d. Linear Least Squares

The classical linear least squares problem may be formulated in probabilistic terms as follows. Let

$$\mathbf{z}=\mathbf{Hx}+\boldsymbol{\nu}$$

where $\mathbf{z}$ is an n-dimensional observation vector, $\mathbf{x}$ a p-dimensional parameter vector, $\mathbf{H}$ an $n\times p$ dimensional matrix of independent variables, and $\boldsymbol{\nu}$ a mean zero random vector with positive definite covariance matrix $\boldsymbol{\Sigma}$. Then

$$\mathcal{E}\mathbf{z}=\mathbf{Hx}$$

and the covariance matrix of $\mathbf{z}$ is $\boldsymbol{\Sigma}$,

$$\mathcal{E}(\mathbf{z}-\mathbf{Hx})(\mathbf{z}-\mathbf{Hx})'=\boldsymbol{\Sigma} \ .$$

We call $\mathbf{Hx}$ the *regression function* of $\mathbf{z}$ on $\mathbf{x}$ and call $\boldsymbol{\nu}$ the *residual*. It will be assumed that $n \geqq p$ and that $\rho\mathbf{H}=p$.

We wish to find the "best" estimate of the parameter vector $\mathbf{x}$ based on the observations $\mathbf{z}$. To be more precise, let $\boldsymbol{\gamma}$ be a p-dimensional vector. We wish to find an n-dimensional vector $\mathbf{g}$ such that $\mathbf{g}'\mathbf{z}$ is a minimum variance unbiased estimate of $\boldsymbol{\gamma}'\mathbf{x}$. This is our definition of "best". Since $\mathbf{g}'\mathbf{z}$ is a linear combination of the observations $\mathbf{z}'=\{z_1,\cdots,z_n\}$, we see that $\mathbf{g}'\mathbf{z}$ is a *linear* estimate. We call such an estimate a best linear unbiased estimate (BLUE).

Since we want $\mathbf{g}'\mathbf{z}$ to be unbiased, we must have

$$\mathcal{E}\mathbf{g}'\mathbf{z}=\mathbf{g}'\mathbf{Hx}=\boldsymbol{\gamma}'\mathbf{x} \ .$$

This implies that $\mathbf{g}'\mathbf{H}=\boldsymbol{\gamma}'$. Now we must minimize $\text{Var}\,\mathbf{g}'\mathbf{z}=\mathbf{g}'\boldsymbol{\Sigma}\mathbf{g}$ subject to the constraint $\mathbf{g}'\mathbf{H}=\boldsymbol{\gamma}'$. If we introduce the vector Lagrange multiplier $\boldsymbol{\lambda}$ (see Appendix B), then we have the constrained minimization problem

$$v=\mathbf{g}'\boldsymbol{\Sigma}\mathbf{g}-2\mathbf{g}'\mathbf{H}\boldsymbol{\lambda} \ .$$

Differentiating the scalar v with respect to the vector $\mathbf{g}$ and then setting $dv/d\mathbf{g}$ equal to the n-dimensional zero vector leads to

$$\mathbf{g}=\boldsymbol{\Sigma}^{-1}\mathbf{H}\boldsymbol{\lambda} \ . \tag{1}$$

Thus $\mathbf{g}'\mathbf{H}=\boldsymbol{\gamma}'$ implies

$$\boldsymbol{\lambda}=(\mathbf{H}'\boldsymbol{\Sigma}^{-1}\mathbf{H})^{-1}\boldsymbol{\gamma} \tag{2}$$

and from (1)

$$\mathbf{g}=\boldsymbol{\Sigma}^{-1}\mathbf{H}(\mathbf{H}'\boldsymbol{\Sigma}^{-1}\mathbf{H})^{-1}\boldsymbol{\gamma} \ . \tag{3}$$

Therefore

$$\mathbf{g}'\mathbf{z}=\boldsymbol{\gamma}'(\mathbf{H}'\boldsymbol{\Sigma}^{-1}\mathbf{H})^{-1}\mathbf{H}'\boldsymbol{\Sigma}^{-1}\mathbf{z} \tag{4}$$

is the BLUE of $\boldsymbol{\gamma}'\mathbf{x}$.

Now suppose that $\hat{x}_k$ is the BLUE of x_k, the kth component of $\mathbf{x}$. Then we call $\hat{\mathbf{x}}'=\{\hat{x}_1, \cdots ,\hat{x}_p\}$ the Markov estimate (ME) of $\mathbf{x}$. It is not difficult to show that

$$\hat{\mathbf{x}}=(\mathbf{H}'\boldsymbol{\Sigma}^{-1}\mathbf{H})^{-1}\mathbf{H}'\boldsymbol{\Sigma}^{-1}\mathbf{z} \ . \tag{5}$$

As a consequence of this formula $\boldsymbol{\gamma}'\hat{\mathbf{x}}$ is the BLUE of $\boldsymbol{\gamma}'\mathbf{x}$.

We see that $\hat{\mathbf{x}}$ is an unbiased estimate of $\mathbf{x}$ since from (5), $\mathcal{E}\mathbf{z}=\mathbf{H}\mathbf{x}$ implies

$$\mathcal{E}\hat{\mathbf{x}}=(\mathbf{H}'\boldsymbol{\Sigma}^{-1}\mathbf{H})^{-1}\mathbf{H}'\boldsymbol{\Sigma}^{-1}\mathbf{H}\mathbf{x}=\mathbf{x} \ .$$

The covariance matrix $\mathbf{P}$ of $\hat{\mathbf{x}}$ is

$$\begin{aligned}\mathbf{P}&=\mathcal{E}(\hat{\mathbf{x}}-\mathbf{x})(\hat{\mathbf{x}}-\mathbf{x})'\\&=(\mathbf{H}'\boldsymbol{\Sigma}^{-1}\mathbf{H})^{-1}\mathbf{H}'\boldsymbol{\Sigma}^{-1}\mathcal{E}[(\mathbf{z}-\mathbf{H}\mathbf{x})(\mathbf{z}-\mathbf{H}\mathbf{x})']\boldsymbol{\Sigma}^{-1}\mathbf{H}(\mathbf{H}'\boldsymbol{\Sigma}^{-1}\mathbf{H})^{-1}\\&=(\mathbf{H}'\boldsymbol{\Sigma}^{-1}\mathbf{H})^{-1} \ ,\end{aligned} \tag{6}$$

since $\boldsymbol{\Sigma}$ is the covariance matrix of $\mathbf{z}$, and

$$\operatorname{Var} \boldsymbol{\gamma}'\hat{\mathbf{x}}=\boldsymbol{\gamma}'(\mathbf{H}'\boldsymbol{\Sigma}^{-1}\mathbf{H})^{-1}\boldsymbol{\gamma}=\boldsymbol{\gamma}'\mathbf{P}\boldsymbol{\gamma} \ . \tag{7}$$

In Section 1 we considered the least squares fit

$$\hat{\hat{\mathbf{x}}}=(\mathbf{H}'\mathbf{H})^{-1}\mathbf{H}'\mathbf{z} \tag{8}$$

[where we have replaced $\hat{\mathbf{x}}$ in (1.11) by $\hat{\hat{\mathbf{x}}}$]. In probabilistic language we call $\hat{\hat{\mathbf{x}}}$ the least squares estimate (LSE) of $\mathbf{x}$. It is unbiased ($\mathcal{E}\hat{\hat{\mathbf{x}}}=\mathbf{x}$) and has covariance matrix

$$\mathbf{Q} = \mathcal{E}(\hat{\hat{\mathbf{x}}} - \mathbf{x})(\hat{\hat{\mathbf{x}}} - \mathbf{x})' = (\mathbf{H}'\mathbf{H})^{-1}\mathbf{H}'\mathbf{\Sigma}\mathbf{H}(\mathbf{H}'\mathbf{H})^{-1} \ . \tag{9}$$

Since $\hat{\mathbf{x}}$ is the ME of $\mathbf{x}$, we must have $\mathbf{Q} - \mathbf{P}$ nonnegative definite.

Now suppose $\mathbf{\Sigma} = \sigma^2\mathbf{I}$ where σ^2 is a positive constant. Then

$$\hat{\mathbf{x}} = \hat{\hat{\mathbf{x}}} \tag{10}$$

[see (5) and (8)] and

$$\mathbf{P} = \mathbf{Q} = \sigma^2(\mathbf{H}'\mathbf{H})^{-1}$$

[see (6) and (9)]. Also,

$$\text{Var}\,\boldsymbol{\gamma}'\,\hat{\mathbf{x}} = \text{Var}\,\boldsymbol{\gamma}'\,\hat{\hat{\mathbf{x}}} = \sigma^2\boldsymbol{\gamma}'(\mathbf{H}'\mathbf{H})^{-1}\boldsymbol{\gamma} \ . \tag{11}$$

The fact that the LSE is the ME when the covariance matrix of the observations is a scalar times the identity matrix [see (10)] is the essential content of the Gauss-Markov theorem.

APPENDIX *B*

Function Theory

a. Lagrange Multipliers

Let $\mathbf{x}' = \{x_1, \cdots, x_n\}$ be an n-dimensional vector, and let

$$y = f(\mathbf{x}) \tag{1}$$

be a differentiable function of $\mathbf{x}$ for $\mathbf{x}$ in some open set $\mathfrak{D}$ of n-dimensional Euclidean space. We can, in general, using elementary calculus, find those values of $\mathbf{x}$ for which the differential dy vanishes. In certain applications, however, it is necessary to find the stationary points of (1) subject to constraints of the form

$$\boldsymbol{\phi} = \mathbf{0} \tag{2}$$

where $\boldsymbol{\phi}' = \{\phi_1(\mathbf{x}), \cdots, \phi_m(\mathbf{x})\}$ and the $\phi_i(\mathbf{x})$ are differentiable functions of $\mathbf{x}$ for $\mathbf{x} \in \mathfrak{D}$. Such maximum-minimum problems may be treated using the *Lagrange method of undetermined multipliers*.

Suppose, then, we wish to find the values of $\mathbf{x}$ subject to the conditions (2) for which the differential of y vanishes. We shall assume that $m < n$, for otherwise (2) would determine $\mathbf{x}$. From (1)

$$dy = \frac{\partial y}{\partial \mathbf{x}'} d\mathbf{x} \tag{3}$$

where

$$\frac{\partial y}{\partial \mathbf{x}'} = \left\{ \frac{\partial y}{\partial x_1}, \cdots, \frac{\partial y}{\partial x_n} \right\}$$

and

$$(d\mathbf{x})' = \{dx_1, \cdots, dx_n\} \ .$$

Now if $m=0$, our problem becomes the ordinary maximum-minimum problem of the calculus. Since in this case the components of $d\mathbf{x}$ are independent variables, a necessary condition for a maximum or minimum is that the differential coefficients $\partial y/\partial x_i$, $1 \leqq i \leqq n$, vanish. Presumably from the n equations

$$\frac{\partial y}{\partial \mathbf{x}} = \mathbf{0} \tag{4}$$

we may solve for those values of $\mathbf{x}$ which satisfy (4) and hence which make dy vanish.

Our problem is not quite so simple. From (2)

$$\frac{\partial \boldsymbol{\phi}}{\partial \mathbf{x}'} \, d\mathbf{x} = \mathbf{0} \ . \tag{5}$$

These m equations impose conditions on $d\mathbf{x}$ and reduce the number of independent variables from n to $n-m$. If we multiply the ith equation of (5) by λ_i [where λ_i is an arbitrary parameter (the Lagrange multiplier) to be determined later] and then add them to (3) we obtain

$$\left(\frac{\partial y}{\partial \mathbf{x}'} + \boldsymbol{\lambda}'\frac{\partial \boldsymbol{\phi}}{\partial \mathbf{x}'}\right) d\mathbf{x} = 0 \tag{6}$$

where

$$\boldsymbol{\lambda}' = \{\lambda_1, \cdots, \lambda_m\} \ .$$

Of course only $n-m$ of the n variables dx_i are independent. However, we have m of the λ_i at our disposal. Now $(n-m)+m=n$; hence we may set the coefficients of $dx_1, \cdots, dx_n$ equal to zero where $n-m$ of these coefficients vanish by virtue of the independence of $n-m$ of the dx_i, and the remaining coefficients vanish by an appropriate choice of the λ_i. We therefore have the n equations

$$\frac{\partial y}{\partial \mathbf{x}'} + \boldsymbol{\lambda}'\frac{\partial \boldsymbol{\phi}}{\partial \mathbf{x}'} = \mathbf{0}'$$

from which we may solve for $\mathbf{x}$ in terms of $\boldsymbol{\lambda}$. Then $\boldsymbol{\lambda}$ may in turn be determined from (2).

b. The Weierstrass M-test

If an infinite series of numbers converges, there is no guarantee that a rearrangement of the series will converge, let alone to the same value.

Also, if an infinite series of functions defined and continuous on a common interval converges, there is no guarantee that the sum converges to a continuous function. There are, however, additional, sufficient conditions that we may impose on infinite series that help avoid such unpleasant phenomena. Roughly speaking, if an infinite series converges *absolutely* we may apply arithmetic operations to the series just as if it were a finite sum; and if it converges *uniformly* we may apply the operations of the calculus to the series just as if it were a finite sum.

If

$$u_1(x) + u_2(x) + \cdots + u_n(x) + \cdots \tag{1}$$

is a series of functions defined on a common interval I, then we say (1) converges absolutely if

$$|u_1(x)| + |u_2(x)| + \cdots + |u_n(x)| + \cdots$$

converges. If there exists a function $u(x)$ such that given any $\epsilon > 0$ there exists an N such that for all $n > N$ and all $x \in I$,

$$|s_n(x) - u(x)| < \epsilon$$

where $s_n(x)$ is the nth partial sum of (1); then we say (1) converges uniformly to $u(x)$. For example, if (1) converges uniformly and each $u_n(x)$ is continuous on I, then so is $u(x)$, and

$$\sum_{n=1}^{\infty} \int_I u_n(x)dx = \int_I \sum_{n=1}^{\infty} u_n(x)dx \ .$$

A useful convenient sufficient criterion to determine whether a series of functions converges absolutely and uniformly is the Weierstrass M-test (see [Miller, 1957].) It states that if

$$M_1 + M_2 + \cdots + M_n + \cdots$$

is a convergent series of nonnegative numbers, and if

$$|u_n(x)| \leqq M_n$$

for all n and all $x \in I$ then (1) converges absolutely and uniformly on I.

To prove this result let $\epsilon > 0$ be assigned and choose an N such that for all $n > N$ and all $p > 0$,

$$M_{n+1} + \cdots + M_{n+p} < \epsilon \ .$$

Then for $n > N$,

$$|u_{n+1}(x) + \cdots + u_{n+p}(x)| \leqq |u_{n+1}(x)| + \cdots + |u_{n+p}(x)|$$
$$\leqq M_{n+1} + \cdots + M_{n+p} < \epsilon \ .$$

Now apply the Cauchy criterion.

c. The Riemann-Lebesgue Lemma

If $[a,b]$ is a finite interval then we easily can evaluate

$$I_N = \int_a^b \sin Nx \, dx \ .$$

Trivially,

$$I_N = \frac{\cos Na - \cos Nb}{N} \ ,$$

and as N increases without limit, I_N approaches zero. Now suppose $f(x)$ is a smooth function. Then over a small subinterval $[\alpha,\beta]$ of $[a,b]$ the function $f(x)$ is practically a constant, say C. Thus

$$I'_N = \int_\alpha^\beta f(x) \sin Nx \, dx \approx C \int_\alpha^\beta \sin Nx \, dx$$

and I'_N is small for N large. This is the essential idea behind the proof of the Riemann-Lebesgue lemma. Rigorizing the above qualitative remarks leads to:

Riemann-Lebesgue Lemma. Let $f(x)$ be integrable on $[a,b]$. Then

$$\lim_{N\to\infty} \int_a^b f(x) \sin Nx \, dx = 0 \ .$$

The most useful variation of the Lemma is given by the corollary below, for which we give a formal proof.

Corollary. Let $f(x)$ be integrable on $[a,b]$ and let ξ be an interior point of the interval. Then if $f(x)$ has a right- and left-hand derivative at $x = \xi$,

$$\lim_{N\to\infty} \int_a^b f(x) \frac{\sin N(x-\xi)}{x-\xi} \, dx = \frac{\pi}{2} [f(\xi+) + f(\xi-)] \ . \qquad (1)$$

Proof. Write

$$\int_a^b f(x)\,\frac{\sin N(x-\xi)}{x-\xi}\,dx=\int_a^\xi f(x)\,\frac{\sin N(x-\xi)}{x-\xi}\,dx+\int_\xi^b f(x)\,\frac{\sin N(x-\xi)}{x-\xi}\,dx$$

and make the change of variable $u=\xi-x$ in the first integral on the right, and $u=x-\xi$ in the second. Then

$$\int_a^b f(x)\,\frac{\sin N(x-\xi)}{x-\xi}\,dx=\int_0^{\xi-a} f(\xi-u)\,\frac{\sin Nu}{u}\,du + \int_0^{b-\xi} f(\xi+u)\,\frac{\sin Nu}{u}\,du\ . \tag{2}$$

Now write the first integral on the right-hand side of (2) as

$$\int_0^{\xi-a} f(\xi-)\,\frac{\sin Nu}{u}\,du+\int_0^{\xi-a}\frac{f(\xi-u)-f(\xi-)}{u}\,\sin Nu\,du\ . \tag{3}$$

By hypothesis the left-hand derivative of $f(x)$ exists at $x=\xi$. Hence by the Riemann-Lebesgue lemma,

$$\lim_{N\to\infty}\int_0^{\xi-a}\frac{f(\xi-u)-f(\xi-)}{u}\,\sin Nu\,du=0\ .$$

In the first integral in (3) make the change of variable $v=Nu$. Then

$$\int_0^{\xi-a} f(\xi-)\,\frac{\sin Nu}{u}\,du=f(\xi-)\int_0^{N(\xi-a)}\frac{\sin v}{v}\,dv\ .$$

But

$$\int_0^\infty \frac{\sin v}{v}\,dv=\frac{\pi}{2}\ .$$

Thus [see (2)]

$$\lim_{N\to\infty}\int_0^{\xi-a} f(\xi-u)\,\frac{\sin Nu}{u}\,du=\frac{\pi}{2}\,f(\xi-)\ .$$

Similarly, since $f(x)$ has a right-hand derivative at $x=\xi$, we have from the second integral on the right-hand side of (2),

$$\lim_{N\to\infty}\int_0^{b-\xi} f(\xi+u)\frac{\sin Nu}{u}\,du=\frac{\pi}{2}f(\xi+)\ .$$

These last two equations establish (1). ■

d. Residue Theory

Let $P(z)$ and $Q(z)$ be relatively prime polynomials in the complex variable z. Let $z_1, \cdots, z_q$ be the zeros of $Q(z)$. Then we may write

$$Q(z)=a \prod_{k=1}^{q} (z-z_k) \ , \ a \neq 0 \ .$$

We shall assume that the z_k are distinct, that is, that Q has simple zeros. Then if P is a polynomial of degree less than q, we may expand the rational function

$$f(z)=\frac{P(z)}{Q(z)}$$

in partial fractions, viz.:

$$f(z)=\sum_{k=1}^{q} \frac{\sigma_k}{z-z_k}$$

where

$$\sigma_k=\frac{P(z_k)}{Q'(z_k)} \ , \ 1 \leqq k \leqq q \ .$$

Thus if Γ is a negatively sensed simple closed smooth curve and $z_{k_1}, \cdots, z_{k_n}$ are interior to the bounded region A bounded by Γ, while $z_{k_{n+1}}, \cdots, z_{k_q}$ are exterior to A, (see Fig. 4) then

$$\frac{1}{2\pi i} \int_\Gamma f(z)dz = \sum_{\alpha=1}^{n} \sigma_{k_\alpha}$$

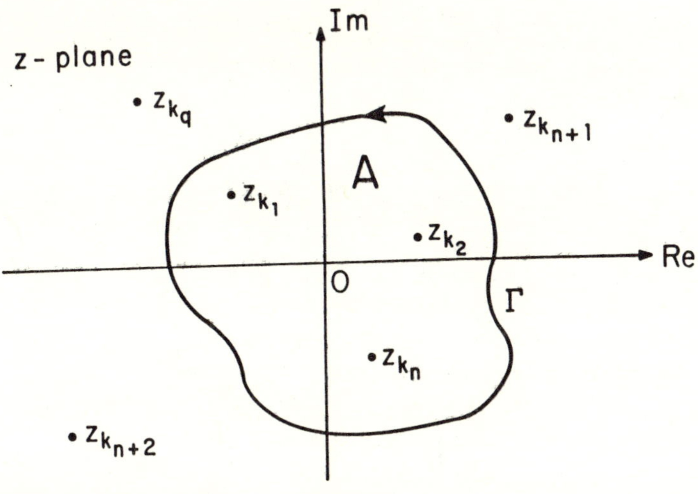

Figure 4

since

$$\int_\Gamma \frac{dz}{z-z_k} = 2\pi i \quad \textit{if} \quad z_k \in A$$

$$= 0 \qquad \textit{if} \quad z_k \notin A \ .$$

We call σ_k the residue of f at $z=z_k$. What we have just stated is a special case of the famous *Cauchy integral formula* (see [Miller, 1970]). However, it is sufficiently general for our purposes.

For example, let

$$P(z) = z^2$$

$$Q(z) = z^4 + z^2 + 1 \ .$$

Then the zeros of $Q(z)$ are

$$z_1 = \alpha$$

$$z_2 = \overline{\alpha}$$

$$z_3 = -\alpha$$

$$z_4 = -\overline{\alpha}$$

where

$$\alpha = \tfrac{1}{2}(1 + i\sqrt{3}) \ .$$

Then if Γ is the curve illustrated below (Fig. 5),

$$\frac{1}{2\pi i} \int_\Gamma \frac{P(z)}{Q(z)} \, dz = \sigma_1 + \sigma_4 \ .$$

Since $Q'(z) = 4z^3 + 2z$,

$$\sigma_1 = \frac{\alpha}{4\alpha^2 + 2}$$

and

$$\sigma_4 = -\frac{\overline{\alpha}}{4\overline{\alpha}^2 + 2} \ .$$

Some simple algebra now shows that

$$\frac{1}{2\pi i} \int_\Gamma \frac{z^2}{z^4 + z^2 + 1} \, dz = \frac{1}{2i\sqrt{3}} \ .$$

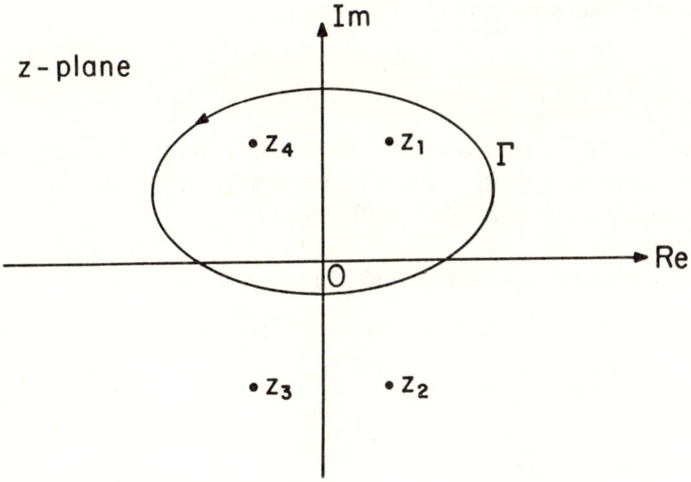

Figure 5

e. Linear Difference Equations

Let y be a function whose domain is the integers. Then

$$\mathcal{L}y \equiv a_0 y(x+2)+a_1 y(x+1)+a_2 y(x)=0 \ , \ a_0 a_2 \neq 0 \tag{1}$$

is called a second order homogeneous *linear difference equation* (or recursion relation). If a_0, a_1, a_2 are (real) constants, then it is easy to find solutions of (1). Namely, let $y(x)=\lambda^x$ where λ is an arbitrary constant. Then

$$\mathcal{L}\,\lambda^x=(a_0\lambda^2+a_1\lambda+a_2)\lambda^x$$

and if $\mathcal{L}\,\lambda^x$ is to be zero, λ must be a root of the *indicial equation*

$$a_0\lambda^2+a_1\lambda+a_2=0 \ . \tag{2}$$

The reader recognizes that the approach is analogous to that employed in finding solutions of linear *differential* equations with constant coefficients.

Since (2) has two roots (neither of which can be zero since $a_2 \neq 0$) we obtain two solutions of (1). Suppose λ is a root of (2). Then if λ is positive, λ^x is a solution of (1). If λ is negative, then λ^x is still a solution of (1), although it is sometimes convenient to write the solution in the form $|\lambda|^x \cos \pi x$. If (2) has complex roots, say $|\lambda|e^{i\theta}$ and $|\lambda|e^{-i\theta}$, the solutions corresponding to these roots are $|\lambda|^x \cos \theta x$ and $|\lambda|^x \sin \theta x$. If λ is a root of multiplicity two, then λ^x and $x\lambda^x$ are two solutions of (1).

As in differential equation theory, we associate initial conditions (or boundary conditions) with (1), for example

$$\begin{aligned} y(0) &= \alpha \\ y(1) &= \beta \end{aligned} \tag{3}$$

where α and β are given constants. Then (1) and (3) constitute a *difference system*, and there exists a unique function $\phi(x)$ which satisfies (1) and takes on the prescribed initial conditions of (3).

For example, consider the difference system

$$y(x+2) - 4y(x+1) + 4y(x) = 0 \tag{4a}$$

$$\begin{aligned} y(0) &= \alpha \\ y(1) &= \beta \ . \end{aligned} \tag{4b}$$

The indicial equation is

$$\lambda^2 - 4\lambda + 4 = 0$$

whose roots are both 2. Thus

$$y(x) = A2^x + Bx2^x$$

satisfies (4a) for any constants A and B. To determine A and B we use (4b):

$$\begin{aligned} \alpha &= A \\ \beta &= 2A + 2B \ . \end{aligned}$$

Then solving for A and B,

$$\phi(x) = \alpha 2^x + (\tfrac{1}{2}\beta - \alpha)x2^x$$

is the unique solution of (4).

For further details and generalizations see [Miller, 1968].

References

A. C. AITKEN, *Determinants and Matrices*, Interscience Pub., Inc., New York, 1942

T. W. ANDERSON, *An Introduction to Multivariate Statistical Analysis*, John Wiley, New York, 1958

T. W. ANDERSON, *The Statistical Analysis of Time Series*, John Wiley, New York, 1971

G. BIRKHOFF and S. MacLANE, *A Survey of Modern Algebra*, Macmillan, New York, 1953

J. W. BREWER, *The derivative of the Riccati matrix with respect to a matrix*, IEEE Trans., *AC-22*, 980–983, (1977)

H. CRAMÉR, *Mathematical Methods of Statistics*, Princeton Univ. Press, Princeton, 1946

F. J. H. DON, *The expectation of products of quadratic forms in normal variables*, Statist. Neerlandica, *33*, 73–79, (1979)

J. L. DOOB, *Stochastic Processes*, John Wiley, New York, 1953

P. S. DWYER and F. V. WAUGH, *On errors in matrix inversion*, J. Amer. Statist. Assoc., *48*, 289–319, (1953)

D. K. FADDEEV and V. N. FADDEEVA, *Computational Methods of Linear Algebra*, Freeman, San Francisco, 1963

W. FELLER, *An Introduction to Probability Theory and Its Applications*, 3rd ed., Vol I, John Wiley, New York, 1968

U. GRENANDER and G. SZEGÖ, *Toeplitz Forms and their Applications*, Univ. California Press, Berkeley, 1958

P. R. HALMOS, *Finite Dimensional Vector Spaces*, Van Nostrand, Princeton, 1958

H. L. HAMBURGER and M. E. GRIMSHAW, *Linear Transformations*, Cambridge Univ. Press, Cambridge, 1951

F. E. HOHN, *Elementary Matrix Algebra*, Macmillan, New York, 1958

A. S. HOUSEHOLDER, *The Theory of Matrices in Numerical Analysis*, Blaisdell Pub. Co., New York, 1964

H. O. LANCASTER, *Traces and cumulants of quadratic forms in normal variables*, J. Royal Statist. Soc., *16*, 247–254, (1954)

E. L. LEHMANN, *Testing Statistical Hypotheses*, John Wiley, New York, 1959

D. M. LESKIW and K. S. MILLER, *Convergence of polynomial least squares estimators*, Pro. IEEE, *70*, 520–522, (1982)

C. C. MacDUFFEE, *The Theory of Matrices*, Chelsea Pub. Co., New York, 1946

N. MACON and A. SPITZBART, *Inverses of Vandermonde matrices*, Amer. Math. Monthly, *65*, 95–100, (1958)

J. R. MAGNUS, *The moments of products of quadratic forms in normal variables*, Statist. Neerlandica, *32*, 201–210, (1978)

K. S. MILLER, *Engineering Mathematics*, Rinehart and Co., New York, 1956

K. S. MILLER, *Advanced Real Calculus*, Harper and Brothers, New York, 1957

K. S. MILLER, *Elements of Modern Abstract Algebra*, Harper and Brothers, New York, 1958

K. S. MILLER, *An Introduction to the Calculus of Finite Differences and Difference Equations*, Henry Holt, New York, 1960

K. S. MILLER, *Multidimensional Gaussian Distributions*, John Wiley, New York, 1964

K. S. MILLER, *Linear Difference Equations*, W. A. Benjamin, New York, 1968

K. S. MILLER, *Advanced Complex Calculus*, Dover Publications, Inc., New York, 1970

K. S. MILLER, *Vector Stochastic Processes*, Krieger Pub. Co., Huntington, NY, 1980

K. S. MILLER, *On the inverse of the sum of matrices*, Math. Mag., *54*, 67–72, (1981)

T. MUIR, *A Treatise on the Theory of Determinants,* Dover Publications, Inc., New York, 1960

F. PROSCHAN, *Precision of least squares polynomial estimates,* SIAM Rev., *3*, 230–236, (1961)

J. RIORDAN, *Combinatorial Identities*, John Wiley, New York, 1968

J. SHERMAN and W. J. MORRISON, *Adjustment of an inverse matrix corresponding to a change in one element of a given matrix*, Ann. Math. Statist., *21*, 124–127, (1950)

E. C. TITCHMARSH, *The Theory of Functions*, Oxford Univ. Press, London, 1939

B. L. van der WAERDEN, *Modern Algebra*, Vol I, F. Ungar Pub. Co., New York, 1949

L. WEISNER, *Introduction to the Theory of Equations*, Macmillan, New York, 1938

H. S. WILF, *Matrix inversion by the annihilation of rank*, J. Soc. Indust. Appl. Math., *7*, 149–151, (1959)

M. A. WOODBURY, *Inverting modified matrices*, Statistical Research Group, Memo. Rep. no. *42*, Princeton Univ., Princeton, (1950)

INDEX